The Advisers

About Brookings

The Brookings Institution is a private nonprofit organization devoted to research, education, and publication on important issues of domestic and foreign policy. Its principal purpose is to bring knowledge to bear on the current and emerging policy problems facing the American people.

A board of trustees is responsible for general supervision of the Institution and safeguarding of its independence. The president is the chief administrative officer and bears final responsibility for the decision to publish a manuscript as a Brookings book. In reaching this judgment, the president is advised by a panel of expert readers who report in confidence on the quality of the work. Publication of a work signifies that it is deemed a competent treatment worthy of public consideration but does not imply endorsement of conclusions or recommendations. The Institution itself does not take positions on policy issues.

The Advisers
Scientists in the Policy Process

Bruce L. R. Smith

The Brookings Institution
Washington, D.C.

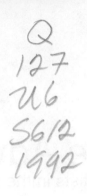

Copyright © 1992
THE BROOKINGS INSTITUTION
1775 Massachusetts Avenue, N.W, Washington, D.C. 20036

Library of Congress Cataloging-in-Publication data:

Smith, Bruce L. R.
    The advisers : scientists in the policy process / Bruce L. R. Smith.
      p.   cm.
    Includes bibliographical references and index.
    ISBN 0-8157-7990-9 (cloth : alk. paper) — ISBN 0-8157-7989-5
(pbk. : alk. paper)
    1. Science and state—United States. 2. Science consultants—
United States. 3. Presidents—United States—Staff. I. Title.
Q127.U6S612   1992
338.97306—dc20   92-6882
                                                  CIP

9 8 7 6 5 4 3 2 1

Preface

THE IDEA FOR THIS BOOK first occurred to me some years ago. I had done a study of think tanks and of analysis in the defense policy process. It seemed a logical next step to generalize the topic into the broader question of science's role in American democracy. The key to understanding modern government lay in the interplay of expert knowledge and politics. For one reason or another I never got the project going or the concept clearly into focus until, in early 1988, I was invited by Joshua Lederberg and David Robinson of the Carnegie Commission on Science, Technology, and Government to undertake a study of the subject. They were interested in science's institutional relations with government, and they particularly wanted me to look at science advisers in the departments and agencies.

An early draft of the book was presented to the Carnegie Commission as a report in December 1989. I gratefully acknowledge the commission's support, and I hope that my efforts have aided the related studies and deliberations under way at the commission. Carnegie colleagues who read drafts, discussed the project extensively, or otherwise provided valuable assistance include David Robinson, Joshua Lederberg, Rodney Nichols, Guyford Stever, William Carey, Mark Schaefer, Jesse Ausubel, David Beckler, William Golden, Maxine Rockoff, and William Perry.

Many Brookings colleagues, as usual, provided encouragement, intellectual stimulation, and valuable critical commentary. I wish to thank in particular Pietro Nivola, Lawrence Korb, Bruce MacLaury, Edwin Dorn, Bert Rockman, Kent Weaver, Robert Katzman, Henry Aaron, William Frenzel, Robert Crandall, Lincoln Gordon, and Peter Malof. Others who were especially helpful were Robert Tarr and William Bonesteel, retired public servants who shared their knowledge of the committee management system with me and made available valuable documentary materials; Brian Murphy of the Administrative Confer-

v

ence of the United States; Richard Berg of Multinational Legal Services, Inc.; Richard Wiley of Wiley, Rein & Fielding; Daniel Fink of Fink Associates; Louis Lanzerotti of AT&T Bell Laboratories; James Wilson of the University of California, Los Angeles; James Carroll of Florida International University; James Smith of the New School of Social Work; Sheila Jasanoff of Cornell University; Donald Elliott of Yale Law School; Klaus Gottstein of the Max Planck Society; Kenneth Pederson of Georgetown University; and Joel Snow of Argonne National Laboratory.

Major Kevin Cunningham of the United States Army read and commented on numerous drafts of chapter 3, generously shared sources and research materials, and stimulated my thinking in many ways. James Dean of the GSA Committee Management Secretariat and his colleagues patiently tutored me on the intricacies of the system. Don K. Price, who introduced me to the study of government and science and whose own work has contributed so much to the field, has continued to be a source of inspiration.

I feel a special debt that I cannot adequately discharge to the many people in the federal government who went out of their way to contribute to this book. The same holds true for the science advisers who discussed with me their agencies, committees, and the business of providing advice. The advisers and the advised, and the men and women who administer the committee system, are the central figures of the story I have tried to tell. I thank them all, and I hope the book makes clear their important contribution to the nation.

Jeffrey Porro and Caroline Lalire edited the manuscript. Tom Hart, Nicole Sanderson, Christianne Millet, Maury Tobin, Stephanie Jones, and Laura Kelly provided research assistance and verification. Laura Walker and her colleagues in the Brookings library were unfailingly courteous and helpful. Marjorie Crow typed numerous drafts. Barbara Kraft prepared the index.

Of course, the views expressed are entirely my own and should not be attributed to the trustees, officers, or other staff members of the Brookings Institution.

BRUCE L. R. SMITH

February 1992
Washington, D.C.

To Courtenay

Contents

Tables

Figures

The Advisers

1 ||| The Political Role of the Science Adviser

THE PRESENCE OF SCIENTISTS as participants and advisers in the nation's policy process is the subject of this book. The focus is the scientific advisory committees in the executive branch. At the time this book was written, in late 1991, there were about 1,000 advisory committees of all kinds in the federal government, reporting to 57 sponsoring agencies.[1] Although no precise classification of these committees is possible, it appears that about half of them are clearly *scientific*—concerned with science, engineering, or research functions. Others could be considered scientific under a looser definition, and most advisory committees attempt to use the methods of science in looking at problems.

These scientific panels can be divided into four broad categories: (1) peer review panels, whose main functions are to review research proposals, award fellowships, or make recommendations on individual grants; (2) program advisory committees, which are charged with strategic oversight of a scientific field or an agency's research and development (R&D) program rather than with the evaluation or award of individual grants or contracts; (3) ad hoc fact-finding or investigating committees, which range from the more specifically technical to those investigating broader problems with important technical dimensions; these committees are typically created for temporary periods to explore a specific problem and to recommend solutions; and (4) the standing committee providing advice to the decisionmaker on broad political-technical issues; this type may be found at the agency level, the bureau or sub-agency level, or the presidential level.

The third and fourth categories—less amenable to tidy classification—are more deeply and directly involved in broad policy matters, and they are the most important committees for the purposes of this study. These committees deal with policy issues that have important scientific or technical dimensions (science *in* policy) rather than with science as such (policy *for* science). This is an important point, for it distinguishes the role of the science adviser in the United States from the adviser's role in other Western democracies, where science advice has been more typically limited to policy for science.

While science advisory committees are the focus of the study, a crisp definition of a scientific committee remains elusive.[2] It is neither obtainable nor necessary for the purposes of this book to have such a definition. Most of the committees studied here are not composed exclusively of scientists, engineers, or research managers. Indeed, many have had a substantial number, and at times even a majority, of non-scientists in their ranks. Similarly it is the rare committee charged with a seemingly nonscientific issue that fails to include its requisite professor, dean, college president, or industrial scientist or engineer.

Most important, the issues at stake—the relationship of specialized knowledge to the general policy context—are the same in the most fundamental sense for scientists as for experts in public finance or taxation, health, national security, transportation, and other areas. The mystique and prestige of science are such that nearly all advisory bodies attempt to cast their findings in terms of empirical evidence based on research and thus label them *scientific* whatever the ostensible subject matter.

Government is of course both politics and administration. The tentative and uneasy compromises, the shifting and unclear boundaries, and the unstable equilibriums between the two domains and outlooks provide the core of the analysis that follows. Science advisers straddle the realms of fact and value, sometimes performing best when addressing a narrow issue with high technical content. At other times they serve the agency and the nation best by acting as a bridge between the ends and the means of policy, by altering prevailing assumptions, and by pointing the way toward new policy paradigms. In all cases the science advisers struggle with the conflicting impulses to be, on the one hand, apolitical and neutral, avoiding the messy, compromise-laden behavior of normal politicians, and to be, on the other, fully engaged on the political battlefield and therefore subject to all the constraints and compromises that affect other political actors.

Key Questions

The book examines a series of advisory committees in action, in an attempt to analyze when and how these committees are useful. My main effort will be to discuss the committees' common problems as well as to describe salient differences in their experiences. Key questions include the following: Do scientists serve the nation best when they function as an "apolitical elite," in Robert Wood's phrase?[3] Or are they most effective when they recognize that they are "in the battle rather than above it?"[4] Do outside advisers merely add a layer of clutter and delay to the policy process? When policy officials do not know what they want or when they fail to obtain what they want from their advisers, is some useful purpose served? Does the network of advisory committees provide useful policy ideas and open up a closed bureaucracy? Does the presence of outside advisers promote accountability by subjecting official thinking to critical scrutiny or detract from accountability by creating shadowy centers of power? Under what circumstances have advisory committees been effective and when not? And in the broadest sense, do advisory committees serve the ends of American democracy? This study attempts to provide some answers by examining a range of advisory committee behavior and drawing on the judgments and opinions of a large number of participants, both those supplying the advice and policymakers who receive and act on or ignore the advice.

Some Preliminary Boundaries

The time period covered is the 1970s and 1980s, broadly the two decades since the 1972 passage of the Federal Advisory Committee Act (FACA). The agencies chosen for consideration span the range from successful advisory operations to mixed or ambiguous cases to outright failures. I start with cases where the advisers—through luck, hard work, a background of political consensus that has facilitated their work, and the presence of able supporting staff people—have managed to render useful service. Then the spotlight shifts to cases where it has been harder, for reasons that will be detailed, for the advisers to find a secure foothold in the tangle of bureaucratic politics, and finally to a case where the participants overwhelmingly agree that few useful results have been achieved. In all cases interviews with advisory committee members as well as with officials from the agencies have been major sources of information and interpretation for my analysis. In addition, of course, I rely on such secondary sources and scholarly works as are available. The participants tell their story in their own words at various places when that usefully illustrates a point.

At all times I base my account on what many participants and observers have told me about the world of giving and receiving advice as well as in some instances on my own direct experience.

Attention is centered on the formal advisory body (the President's Science Advisory Committee, the Defense Science Board, the NASA Advisory Council, the Energy Research Advisory Board, the EPA Science Advisory Board, and the State Department Science Advisory Committee). Usually, however, the central board or committee operates most significantly through panels, subcommittees, or task forces. The relationship of the overall committee to its working parts deserves careful consideration. But I accept the definition that a subgroup is not an *advisory committee* under the FACA (see chapter 2), since it reports to its parent body—the legally constituted advisory committee—and not directly to the agency.

What the FACA calls *utilized committees*, such as those associated with the National Academy of Sciences (with its 800 to 900 separate committees and boards and an annual budget of more than $150 million) or the research institutes operating under contract with government agencies, are similarly excluded. The National Academy of Sciences complex, the American Association for the Advancement of Science, university research centers, and other nongovernmental bodies play important roles in the total science advisory system.[5] But they cannot be treated fully for the purposes of this study. The think tanks that play a modestly important role in the decisionmaking processes of a number of agencies are also excluded. Practical considerations rule out any extensive treatment of consultants and the numerous informal advisers that float around Washington, sometimes alighting with a splash in the news media but more often composing the army of bit players participating at the margins in the great drama of American politics. The one-shot committees—that is, informal committees meeting only once or twice and having no organizational structure—cannot be analyzed either.[6]

The word *formal* excludes, furthermore, the large numbers of informal sources of advice—individual consultants, kitchen cabinets, and so on—that sometimes constitute an important aspect of policy and operations. Indeed some agencies and individual policymakers have chosen to rely heavily on informal mechanisms, or at least on less formal advisory structures, because they do not wish to be burdened by FACA requirements.

I try to show the advisory systems in action to the extent that this is practicable. The intent is not merely a description of structures and institutions but an assessment of how the advisory committees work or fail to work in practice. How, in short, do the advisory mechanisms influence the policymaking process and affect the life of the agency they serve?

In the many interviews for the case studies, a standard set of questions has been asked. The pertinent questions include the following:

— Why was the committee created? What is its mission?

— To whom does the advisory committee report within the agency (that is, who is the client)?

— How is the work assignment chosen? Is it self-initiated, undertaken in response to specific client request, or some mix? Is the committee free to broaden or redefine its mandate as it deems appropriate?

— How are members chosen for service on the committee and how is "fair balance" defined? How and by whom are the officers (chairmen or group leaders) appointed or otherwise designated? What is the role of the chairman or group leader?

— What effects have the conflict-of-interest and open meetings laws (FACA, Freedom of Information Act, and Government in the Sunshine Act) had on the ability to recruit committee members and to conduct effective operations?

—What level of staff support or secretariat is provided for the advisory committee? How important is adequate staff support?

— How active has the committee been? Is workload correlated with effectiveness?

—How formal are the procedures by which recommendations are communicated to the client? Do the recommendations contain suggestions for implementation? In the case of systems involving both formal committees and ad hoc task groups, what are the roles of each? In the case of multilayered systems, what are the roles of the central board as against those of the various sub-units?

— How closely is the advisory committee involved in budgetary issues within the agency?

Judging Success

Advisory systems encompass a wide range of behavior and are as varied in their styles of operations as their respective parent agencies. Thus there can be no uniform model of the "successful" advisory body or system. However, some guidelines do distinguish the more from the less successful activities. A simple rule of thumb is that the adviser or committee, while conforming to the agency's bureaucratic culture, should normally produce some change in attitudes, outlook, and policies, should buy time to allow a process of thought to mature, or should at least reinforce predispositions or policy inclinations beyond what already existed. If nothing changes as a result of the committee's work, one may plausibly infer that the advisory mechanism has served little purpose.

Assessing impact or influence on policy is a highly complex issue. The advisory committee will almost always act in a context in which numerous other influences are present. Internal staff work, external advisers, academic scholars, publicists, and media commentators all will most likely be involved in debating an issue in one forum or another. The advisory committee will often serve as a kind of transmission belt between the wider community of knowledge workers and the officials charged with program and policy responsibilities. James Allen Smith sums up the diffuse impact of expertise in these terms: "The expert rarely contributes the flash of insight that quickly and fundamentally transforms national policy or inspires an innovative law. Instead, experts work slowly, generally building up intellectual capital in a process one scholar characterized as 'knowledge creep.' "[7]

The relations between the internal and external advisers and the relative impact of the advisory committee working with the agency as opposed to the group that seeks influence through publicity and media attention are especially complex. The advisory committees studied in this book are of the insider category, those who work at the behest of and closely with policymakers in the government. They have access to information and the timing of decisions that outsiders do not have, and they view their role as working for the agency. Hence they give up a measure of independence. Yet almost all formal work products of advisory committees, such as reports and studies, eventually become public documents and indeed often are intended to attract media attention and comment. Advisers know that publicity is one tool they have at their disposal, and they can on occasion maneuver the agency into having to respond to unfavorable publicity. But wrangles in the media can have an ephemeral and marginal impact, and experienced advisers play the publicity card with great caution and in close consultation with their agency patrons if they wish to stay in an influential role.

Moreover, a normal ebb and flow is common in the fortunes of all advisory committees with a continued existence. Circumstance, personalities, and shifting political currents all have a bearing on what the committee does and whether anyone notices or cares. For example, consider simple changes in personal relationships that affect how a board or committee operates. The committee that has functioned well under one chairman who has a close personal relationship to the agency head will normally experience a setback when either of those people leaves. Similarly a committee that has a reputation as a vigorous advocate of one set of program goals will often find itself out in the cold when the agency policy changes or when events shift attention in a new direction. The higher one proceeds in the decision hierarchy, and especially when the White House is directly involved, the more the intangible factors, including personal relationships and status considerations, will dominate. The shifting nuances of high politics will influence and inevitably constrain the scope, mandate, and ultimate result of the advisory effort.

An American Phenomenon

No other Western democracy relies as heavily as does the United States on the use of outside advisers. Britain, Canada, and Australia have made use of the Royal Commission, but this mechanism has been used much more sparingly than American advisory committees and has tended to operate within a carefully defined mandate and in close accord with official operating procedures.[8]

There are a number of potential explanations for the American advisory committee phenomenon. First, science and technology, by introducing a new dimension of complexity into public affairs, have wrought a transformation in government structures and policymaking styles. America, as the nation at the forefront of the scientific revolution, exhibits the most widespread and pervasive use of the science adviser.

The growing role of science and technology in public affairs also helps to account for the substantial number of Western democracies after World War II that created science advisory councils of various kinds.[9] But the scientific advisory councils outside the United States have rarely been used for science-*in*-policy issues as opposed to policy-*for*-science concerns.[10]

A second explanation is therefore required to account for national differences in the use of advisers. Constitutional-legal factors create incentives for or against the use of outside experts. The policymaking processes of most Western democracies are typically more executive-centered and less open than the disorderly and loose-jointed pluralism of America. This is not to say that citizen and interest group involvement is absent from the administrative processes of other Western democracies. Citizen and interest group access instead occurs within a corporatist framework; a small number of peak associations participate formally in the administrative process.[11] Parliamentary systems make it much more difficult for citizens, interested experts, or legislative committees to gain access routinely to the inner deliberations of administrative agencies. The existence of an independent legislature in the United States creates opportunities for oversight investigations, devices for prior consultation, and other interventions at many points in the policymaking processes. A small number of perk associations reporting to a few senior officials never came to dominate the more open and disorderly policy process of the United States deriving from our separation of powers.

To the general constitutional framework as an explanatory factor must be added the nation's bureaucratic culture. The American civil service is more open than European systems to outside midcareer entry and more populated with technical specialists in its upper ranks who naturally look to their colleagues on the outside for advice. The American civil service is also less dominated by recruits from a select group of academic institutions. The policy integration functions, performed in the European tradition by an elite administrative class, are weakly developed in the American federal bureaucracy. As a result U.S. government agencies often look to advisory groups for their main policy ideas.

Even in agencies with strong leadership and well-developed internal planning units, the outside advisers may be deemed necessary to validate internal decisions or to act as a link to an important constituency. Executives with a clear policy agenda often find it useful to reinforce their thinking with the help of outsiders. They use advisory committees for a wide range of activities: sending up trial balloons, engineering consent or support among affected constituencies, educating the public, overcoming internal agency opposition to a course of action, delaying action or evading responsibility for actions taken, reassuring the public that high-level attention is being given to an issue, and broadening the base of participation in government.[12]

Beyond the factor of bureaucratic culture the nation's overall organization for science is significant. In most European democracies a large central department of science and industry or ministry of research exists. This ministry is responsible for research funding, links between government labs and industry, and large-scale research activities (in some cases higher education is included as well). This structure gives scientists more weight in decisions on resource allocations but diminishes science's broader influence in policymaking on matters beyond the narrow concerns of the scientific community. In the United States there is no central science department and no central R&D budget. The responsibility for research support rests with the large mission departments that support the research they deem appropriate to advance their program goals. This guarantees that every major department will have some involvement with science and will seek advice on the narrower aspects of research support as well as on the potential application of science to new missions and areas of program development.

Finally, it is necessary to return to the broad political culture for a complete view of the advisory committee phenomenon. The appeal to science has been the coinage of political legitimacy in the nation since we shed the theocracies of the colonial era. And it has been a democratic science rather than an elitist or hierarchical science that has appealed to Americans, who similarly prefer pluralism in politics and denominationalism in religion. America is strongly imbued with an antistatist and antielitist mentality; there is no concept of sovereignty in the American political tradition except for a vague notion of popular sovereignty. All this means a diffusion of authority under the loose banner of science and reason, along with a populist-participatory impulse in policy formation. As a stamp of legitimacy, any serious policy innovation therefore seems to require a wide involvement of experts of all kinds, the affected interests, and citizens. As one advisory committee chairman observed, "We have so many advisory committees because we are a free-market, private sector society. We don't believe in bureaucrats running everything." Another experienced participant-observer remarked, "You have to have five blue ribbon committees all in broad agreement on the same problem before anyone takes you seriously. This is how you manage to get on the agenda of public action in our country."

That advisory groups are made up at least in part of scientists or other experts is a natural concomitant to this line of thinking. The large number of advisory committees in the United States, then, ultimately reflects our rejection of revealed truth and our embrace of practical reason as the basis for political authority. The Madisonian scheme of politics has meant that many partial and competing truths would be the broad impulse behind policymaking, as balanced by the arts of political compromise and the rule of law. In looking at science advisory committees, in short, one deals in microcosm with the larger workings and the animating principles of American democracy. If advisory committees have to struggle to gain attention and to be taken seriously, this is because all political actors have to do the same. Advisory committees are not extrinsic to but an intrinsic part of the pluralist struggle.

This study chronicles the special complexities, challenges, frustrations, and satisfactions of the advisers' role, in the hope of achieving a deeper understanding of the functions scientists perform in America's governing system. But before I turn to the case studies showing the advisory committees in action, the framework of law and regulation within which the committees operate needs to be examined.

2 || The Changing Legal and Institutional Framework for Science Advice

SCIENTIFIC AND TECHNICAL ADVICE in the federal government dates from the early days of the republic. The Founders, people of broad learning and of the Enlightenment, took an active interest in the science of the day and sought to bring technical information to public affairs. The U.S. Constitution, breaking the tradition of authority based on the divine right of kings, established a "new science of politics" at the core of our governing institutions. Madison, Hamilton, Franklin, Jefferson, Adams, Paine, Rittenhouse, and Bartlow, among others, saw government and legislation as a science, a matter of discerning the principles based on knowledge and reason to guide correct actions.[1] Forces and counter forces, in the fashion of Newtonian physics, would at the same time provide the democracy's equilibrium. Practical reason indeed would be a core value of the whole system (to be balanced with such other core values as the arts of political compromise and the rule of law).

The Founders, Science, and the New Democracy

If the Founders did not always believe that government was quite as simple as a machine run on easily discoverable principles, they at least found it convenient to maintain this view in order to debunk the despotisms and prejudices of Europe. They needed to establish a new basis of authority that did not depend on established truths; and they needed to defeat their enemies at home.

Alexander Hamilton and James Madison, arguing the Federalist cause, depended on a faith in the new science of politics and in the knowledge of principles of good government to defend the idea that democracy could be established in a populous and geographically large

11

state as well as in a small community (as Montesquieu had argued).[2] For them government should be structured on the free choice of enlightened citizens rather than on tradition, emotion, and history. This view was disputed as shallow, superficial, and even dangerous by the Anti-Federalists and others, who opposed the adoption of the Constitution. The opponents felt obliged to rebut the effort to invoke the authority of science. Anti-Federalist Luther Martin wrote that "if the framing and approving of the constitution now offered to our acceptance, is a proof of knowledge in the science of government, I not only admit, but I glory in my ignorance."[3] In England Edmund Burke assailed the experimentalism of religious dissenter and chemist Joseph Priestley, whose thought was so congenial to revolutionaries in both America and France. Burke objected to Priestley's thought on the grounds that it rendered "knowledge . . . worse than ignorance" and threatened to "sacrifice the whole human race to the slightest of . . . experiments" while ignoring history, experience, deeper wisdom, and the ultimate moral basis of all authority.[4]

But the Anti-Federalists and the American expositers of Burkian values of tradition and precedent in public affairs were history's losers. The Founders—who celebrated common sense, plain facts, and man's ability to invent and fashion institutions almost from scratch in accord with simple logic—won out decisively. The basis of political legitimacy in America henceforth largely rested on appeals to the authority of science. The United States has continued to rely on the symbols of scientific reasoning, empiricism, and a neutral and apolitical expertise to legitimate public action.

The Founders, as practical people, did not exalt reason or allow their love of science to become a new religion. Scientists would become no new priestly class or establishment but would be controlled by elected officials close to the people who were accustomed to political compromise. A highly theoretical, all-encompassing world view was not the American conception of science, nor would it be the basis for approaching public problems. Factions—that is, groups of citizens—would propose solutions to concrete problems. The government's bureaus would draw on their own practical knowledge in refining these ideas and proposing specific courses of action. The interaction of many separate interests inside and outside of government would add up to the public interest. No one institution or individual, including even the president, would be the authoritative interpreter of the common good. Americans distrusted the concept of sovereignty in the

British or European sense and displayed a preference for wide partic-ipation in the quest for specific solutions to concrete problems.

Ultimately the judiciary, as guardians of fair process among the elements of a pluralist government, would safeguard individual liber-ties as well as foster political stability. American law would be more fact-based and pragmatic than law in Europe and less bound by rules and authoritative decrees from government or other courts. A small government establishment, operating with only limited powers, would inevitably engage the energies of the wider society in the formulation of public policies. Informal advice from many channels outside the government was therefore part of the scheme of things from the start of the national government.

The scientific ethos, or the experimental method, thus permeated American culture, politics, and society almost from the beginning of the republic. The tradition was manifest in many specific aspects of governance and in the liberal-democratic creed underlying public be-havior. America was the ultimate experimental society, forever in-venting itself and applying practical reason to the affairs of state.

There was the contrasting vision—not quite suppressed, but dis-tinctly the counterpoint to the majority fugue—of the natural rights school of thought. This appealed to the Anti-Federalists and to others who embraced the civic republican outlook based on communitarian values rather than the clash of group interests.[5] At various critical mo-ments in the nation's history—the rise of abolitionism and the Civil War, the Progressive Era, the nation's involvement in the world wars of the twentieth century, the civil rights revolution of the 1960s, and the revolt against statism in the 1980s—the communitarian impulse with its roots in natural rights thinking triumphed over the shallower pragmatic and instrumentalist liberal-democratic creed as the source of great events. Less savory aspects of American public life, almost the mirror image of the rationalism that dominated the majority polit-ical culture, erupted in other episodes that led some observers to iden-tify a "paranoid style" in American politics.[6]

The scientific outlook that underlay America's liberal-democratic state was moreover ambiguous in its core concepts. The quest for sci-entific truth, when mixed with the actual working of politics, often became simplified and diluted so that it had merely the appearance of pursuing rational fact finding in the conduct of affairs. Politicians were more interested in using the prestige and objectivity of scientific in-quiry to legitimate their actions than in insisting on the conformance of public policies to the actual complexities of the latest and most scientifically valid knowledge.[7] The American idea of science as a body of knowledge lending disinterested support to "correct" public policies, while it commanded the support of broad sections of public opinion, often failed to comprehend the contingent and the uncertain in scientific explanation. America was therefore a society highly receptive to science, but its very embrace of the scientific ideal in its public life meant that democratic principles would supersede and modify the pure scientific ideal. Tocqueville was only one foreign observer who wondered whether America's democratic principles would influence the workings of the scientific community in such ways as to impede pure science. Whether America was truly a fertile ground for high science in the light of egalitarian and majoritarian sentiment continued to be the subject of intense debate in the nineteenth century and beyond.

Nor was there a hierarchy within the scientific community to settle disputes arising from its various parts. American science was like American politics and religion: decentralized, pluralistic, and denominational rather than subject to an overall authority.[8] The war of the parts against the whole was a familiar American story. It could be virtually guaranteed that on highly controversial matters the disputants would invoke their scientific advisers in a highly partisan fashion. But for scientists to behave in a partisan fashion would debase the coinage of their expertise and have a corruptive effect on the authority of science as an objective body of knowledge. A paradox thus lay at the heart of science's role—the authority of practical reason would replace established truth in the conduct of the nation's public affairs, but as reason became drawn into politics its objectivity, the basis for its appeal, would erode.

The First Advisory Committees

The first uses of advisory committees seem to have come at the presidential level. If one looks back at the early republic, the cabinet may be said to have served as an internal advisory committee to President Washington (and has continued to have this function even while acquiring quasi-operational responsibilities). One of the earliest instances of the appointment of an outside advisory commission came in 1794 after the Whiskey Rebellion. President Washington appointed a commission to investigate the causes of the revolt by western Pennsylvania farmers against the 1791 federal excise tax on spirits and to attempt to mediate a solution.[9] The president applauded the commission's "firmness" and averred that its report proved "that the means of conciliation have been exhausted."[10] He proceeded to suppress the rebellion by force. The president's appointment of the commission had been accompanied by the simultaneous mobilization of 15,000 militiamen, suggesting that in creating the commission he may have had motives beyond the search for neutral advice from distinguished citizens. Yet the symbolism of appointing a commission before he acted is instructive. In a grave challenge to the authority of the national government, the president felt it necessary to legitimate his actions by first investigating the problem, searching for practical advice from disinterested citizens, and acting on the basis of the facts presented to him.

President Jefferson relied on the appointment of a commission to help devise a solution to the dispute over the Yazoo River land claims and acted on the commission's advice (but congressional opposition blocked the solution until the Supreme Court resolved the matter in the *Fletcher* v. *Peck* case of 1810).[11] The commissioning of Meriwether Lewis and William Clark by President Jefferson to explore the western lands in 1803, and their subsequent report, was perhaps one of the first efforts to tap specifically technical advice from outside the government.[12] The expedition supplied the president with knowledge of the continent, but it was also intended to serve Jefferson's strategy of aiding the commercial interests of American fur trappers.

In the beginning, Congress raised no serious objections to presidential actions to draw on outside advice. However, as presidents increased their reliance on private citizens for advice on public matters, various members of Congress grew uneasy, fearing that presidents could circumvent the normal processes of government. Congress accordingly was ready by 1842 to challenge President Tyler to justify his use of outside advisers. "Under what authority," queried a House resolution of February 7, 1842, and "for what purposes and objects," and "out of what fund" had President Tyler appointed a commission to investigate the New York Customs House?[13] President Tyler's reply drew on article 2, section 3, of the Constitution in asserting a broad presidential authority to inform himself on executive activities and matters on which he would potentially recommend legislative action.

President Tyler also invoked what has remained the standard rationale favored by executive branch decisionmakers for an advisory committee, namely, to provide confidential advice and analysis for the decisionmaker prior to the making of a decision:

> The report of the commission is now wanted, by me, *for my own information*. I do not doubt but that it will contain many suggestions . . . most worthy to be recommended by me to Congress. . . . Whether, when made, I shall deem it best to communicate the entire report to Congress, or otherwise make it public . . . will be for my own decision.[14]

Technical Advice in the Nineteenth Century

In the course of the nineteenth century as the federal government grew in size, the relation between the formal bureaucracy and outside advisers became more complicated, especially in the more technical aspects of policymaking. The Founders tended to view science in the broad sense of organized knowledge of all kinds. Many took an active interest in the progress of various branches of learning. The firsthand acquaintance with scientific affairs by the Founders gradually gave way to a routine dependence by politicians on technical knowledge generated within the executive departments or provided by their immediate staffs. As science and technology progressed, political leaders became less likely to have any detailed grasp of scientific matters or even to aspire to such knowledge.

Science nevertheless was increasingly incorporated into the workings of government. As the federal government became more complex, technical skills were institutionalized within the various agencies and departments. Indeed, the American civil service, as it developed, became one of the most scientifically oriented among the bureaucracies of the modern industrial states.[15] Government technical bureaus could increasingly draw on internal resources to perform missions of an applied research nature, only intermittently needing outside technical help. The universities, or at least the twenty or so research universities, emerged as centers of more fundamental inquiry with the aid of private patronage.[16]

Technology played an important part in the economic expansion following the Civil War. The large-scale, vertically integrated firm depended on the industrial laboratory, which emerged over the 1890–

1920 period, for a continuous stream of inventions. The product-oriented research effort of industry proceeded largely on its own track, and had few direct ties with the research activities of government bureaus or with the universities.

The Twentieth Century

The World War I emergency brought about close government-private sector technical cooperation in the development of war industries previously underdeveloped or nonexistent in America (nitrates, optics, aeronautics, and so on). In the process new advisory relationships were forged, including the creation of the National Research Council within the National Academy of Sciences (NAS). The government sought to tap scientific expertise during the war through an advisory committee chaired by Thomas Alva Edison. This mechanism and much of the remaining wartime collaboration fell into disuse after the war. The military services and some other agencies had nevertheless augmented substantially their technical capacities as a result of World War I.

The Depression saw an abortive effort to create a science advisory committee at the White House level, the beginnings of the modern health research empire with substantial extramural research support, the first use of research contracts with universities under the Works Progress Administration, and a complex assortment of temporary consultative arrangements between government and industry.[17] There was in addition a host of new regulatory mechanisms incorporating advisory committees from the affected industries.

The scientific community during the 1930s largely abandoned its apolitical stance and aloofness toward public affairs in the face of depression, the rise of fascism, and the threat of war.[18] With war clouds gathering and with the well-chronicled scientific discoveries leading to the development of nuclear weapons, the modern era began. It has been characterized by close interdependence between scientists inside and outside government. The wartime developments, reinforced by the postwar tensions, form the background for the current era of government-science interdependence and for this study's interest in the role of science advisers.

The war signified that henceforth the technical efforts of the national security agencies would operate on an almost full mobilization footing, that is, drawing continually on the nation's best technical talents and resources. Formal and informal advisory mechanisms would inevitably play an important part in this process as they had done during the war through the Munitions Board, the Office of Scientific Research and Development, and other entities. As the federal government acquired the responsibility to nurture the universities' basic science efforts, academic scientists themselves were called on to participate in government as evaluators, critics, and advisers. This participation came first in the award of grants, then more broadly in the design of strategies for the growth of science, and finally in the still broader role of assessing the policy significance of scientific and technical developments.[19]

The postwar system thus involved several key features: (1) increased government technical activities; (2) wider use of outside talent as consultants, intermittent advisers, and temporary employees; (3) these outsiders bringing the most up-to-date technical knowledge to policy deliberations and monitoring the quality, performance, and directions of the government's own technical efforts; and (4) these outsiders rendering this service according to informal rules of the game that were not always understood by the participants.

The rules of the game sought to protect both the accountability of government officials and the scientists' independence, but misunderstandings over the responsibilities of each side were common. Yet the urgency and the evident commitment to the national interest by all parties provided a broad base of support for the postwar system. The original understandings loosened as the national security issues became more complex and as new social concerns became the focus of science and technology policy in the late 1960s and the 1970s. The advisory system then increasingly began to reflect the wider conflicts in society and the public's growing doubts that scientists possessed any special wisdom on policy matters.

An inevitable complication of the advisory system, moreover, was the fact that the scientists could hardly escape some degree of self-interest even though they ostensibly sought to provide objective advice to policymakers. In the nineteenth and early twentieth century the scientific community was not dependent on the federal patron. It could more readily be seen as apart from the political fray. As science became dependent on federal research support and as big government drew on the resources of industry for large-scale R&D projects linked to defense, space, atomic energy, and other national purposes, science became the object of public action as well as external critic and adviser to policymakers. The academic scientists who indirectly benefited from the growth of federal funding to their universities found it harder to establish their complete objectivity. Scientists and engineers from industries whose firms might build complex systems on government contracts had even more evident vested interests. Concerns over possible conflicts of interest were therefore bound to emerge as the system matured in the postwar era.

Efforts to Reform the System

In sum, the setting in which federal agencies seek and use technical advice derives from the logic of the nation's liberal-democratic ideology and the separation-of-powers constitutional system. The divided chain of command in which officials report both to their superiors in the executive branch and to congressional committees prevents the tight administrative discipline found in the European parliamentary democracies. The independent judiciary in the United States provides a further avenue for opening up the administrative process to outside influences. The U.S. bureaucratic culture reinforces the features stemming from the constitutional structure. American officials, because they are specialists with close ties to their professional peers outside the government, are more than usually receptive to outside influences. Federal agencies commonly seek both support and counsel from the outside organizations whose backing they need in their internal jockeying for position.

After World War II there arose a striking increase in the complexity of governmental decisionmaking, especially in the agencies heavily dependent on R&D. This added to the pressures to secure the counsel of outside experts and to lay claim to the new program areas arising from technological developments.

Reform efforts in this circumstance would logically be directed toward curbing the excesses of pluralism and strengthening the formal governmental machinery. Scholars of public administration have generally been alert to the potential dangers of "federalism-by-contract,"[20] the contract state,[21] and third-party government.[22] Public administrators concerned with accountability and policy coherence have occasionally generated broad support behind efforts to strengthen central management.[23] Management reforms in the past several decades such as planning-programming-budgeting and zero-based budgeting have sought a strengthening of central executive authority.[24] But the reformers calling for a stronger chief executive have more often than not been voices in the wilderness.

"The cure," as Henry Mencken observed, "for the evils of democracy is more democracy." This Progressive Era watchword found its modern resonance in the major developments affecting America's governing institutions in the 1970s. Congress, the most powerful and most democratic legislative body in the world, made itself more democratic internally. Concomitantly it became a more powerful presence in the internal affairs of the executive branch.

The judiciary for its part has broadened its doctrines of standing to sue, and thus has further enhanced the rights of private litigants.[25] The press and broadcast media in the United States, more vigorous and unfettered than their counterparts in other Western democracies, have become more assertive and aggressive. Judicial decisions have made it all but impossible for a journalist to libel public officials.

The federal bureaucracy—already more open, representative, and accessible to citizens and outside interests than its Western counterparts—has become even more open, more representative of social interests, and more exposed to outside influences. There has also been a greater internal dispersion of power centers resistant to central authority. The interplay of these various pluralizing forces forms the setting for agency use of advisory committees and outside experts in the modern period.

The recent trends have encouraged the widespread use of advisory committees and outside experts and at the same time has created greater obstacles to their effective use. It is not surprising that advisory behavior exhibits in microcosm the broad forces affecting the political system. In the vast subterranean stream of activity that constitutes the "real" work of government, advisory committees, experts, associations, and formal and informal groups struggle to give practical meaning to the broad themes that define the politics of an era.

In recognition of the fact that advice may help to advance one side's broad goals, government has sought to define the rules of the game for advisory activity so as to ensure a reasonably fair and open process. The major political protagonists have tacitly agreed that the system of advice should not be rigged grossly to advance any one set of interests. First by executive order and then by statute, the government has attempted to ensure that outside advisory groups would have fair, open, and reasonably equal access to decisionmakers.

Another motive was to create a set of policies that would guard against conflicts of interest. For inherent in the advisory process was the potential for misusing the access to information and inside thinking to advance one's own special or narrow group interests. Scientific experts, even while striving to be disinterested, are human beings with their own interests and frailties. They are for example employed by universities, firms, or institutes that may be affected by government action. Like other citizens, experts may confuse public and private ends or find a close affinity between what is good for themselves and the larger public good.

Thus, three aspects of the regulatory framework for advisory committees are of particular importance: openness requirements, balance in the membership of advisory committees, and conflict-of-interest issues. These issues can best be brought into focus by first discussing the Federal Advisory Committee Act (FACA) and noting its strengths and weaknesses. As the major law regulating advisory committees, the FACA is startling for its omissions, the circumstances of its birth, and the conflicting assumptions built into House and Senate versions of the bill, which were awkwardly and confusedly fused together. In its essence the FACA sought incompatible goals: on the one hand, to open up the decisionmaking process by bringing in many additional outside advisory groups on the theory that democracy would thereby be best served; on the other, to control the creation of advisory committees on the theory that American government lacked internal discipline and coherence, managerial capacity, and accountability. The result was to impose on the agencies a contradictory regulatory regime that could prove to be burdensome, modestly useful, or largely irrelevant depending upon how it was implemented.

The Adoption of the Federal Advisory Committee Act

The Federal Advisory Committee Act (5 U.S.C. sections 1–15), passed in October 1972, is the major law regulating advisory committees. This act pays explicit attention to the role of federal advisory committees but not to scientific or technical advisory committees as a separate class. The act grew out of a concern that federal practices regarding the use of advisory committees were confused, inconsistent, and in need of clarification both to protect the public interest and to make full and effective use of outside advice. Congressional concerns, in particular, centered on whether advisory committees were balanced fairly in their membership (it was believed that many advisory committees were strongly biased in favor of industry) and whether they operated in an open and transparent process.

Before the late 1950s there was no formal regulation of agency advisory committees despite the widespread use of such bodies dating back to the early days of the republic. The regulation came about as the culmination of several decades of worry over the proper role of outside groups in governmental decisionmaking, especially as the public sector grew in size and importance. The critics of advisory committees were a relatively small group of legislators, usually populists suspicious of big business. They occasionally could muster the ire of enough colleagues to cause the agencies to take notice. Not much happened initially as a result of this concern, but a receptive climate for legislation was gradually created and finally led to action.

The story begins with the New Deal. After passage of the National Industrial Recovery Act of 1933, some members of Congress became concerned over the activities of the committees formed (usually from trade associations) to decide on production codes in each affected industry.[26] In the 1935 *Schecter Poultry Corp.* v. *United States* case a chicken wholesale dealer brought suit and successfully challenged the program as an unconstitutional delegation of legislative powers, establishing the principle that advisory committees cannot exercise governmental powers.[27]

The increased presence of the federal government in the nation's economic affairs continued to prompt congressional concerns. Republicans worried that increasing regulation would stifle individual rights. They sought to ensure that regulatory agencies took account of citizen reactions before imposing unnecessary burdens. Democrats, on the other hand, feared that agency efforts to elicit the views of regulated industries in the name of making regulation more effective would give undue weight to industry groups and would work against the interests of the wider unorganized public.

After the war a new element emerged. The Department of Justice took up the issue of whether standards were needed to regulate agency relations with advisory committees in the interest of averting antitrust violations. Justice in the early 1950s proposed standards having "a remarkable similarity to the basic requirements of the FACA . . . twenty years later."[28]

A few years later congressional hearings were held, and a bill, H.R. 7390, was reported that sought to enact similar standards into law.[29] The bill was aimed at dealing with potential abuses that might result from advisory committees eliciting "the support of the regulated in the process of regulation." In particular such groups representing special interests might threaten to usurp the "managerial functions which are the responsibility of the governmental agency." The bill would "clothe these [advisory] committees with official status and make their proceedings a matter of public record . . . ," enabling Congress to be "kept informed and . . . better able to give intelligent consideration to the possible necessity for further legislation in this area."[30]

The executive agencies objected vigorously to the proposed legislation, insisting that potential abuses could be better and more flexibly handled by the agencies themselves. The House nonetheless passed the measure with minor amendments on June 10, 1957. It failed to pass the Senate. The issue thereupon moved to the legislative back burner, where it continued to attract modest attention on occasion. But at the time the traditional argument prevailed: such internal administrative matters were best left to the executive branch. The Senate acted as a deliberative chamber checking the not very bold impulses of the House.

Meanwhile partly to stave off unwanted congressional action limiting its flexibility, the executive branch began to consider steps of its own to regulate advisory practices. In 1959 the Bureau of the Budget (BOB) issued the bulletin *Standards and Procedures for the Utilization of Public Advisory Commissions by Government Departments*.[31] The standards and procedures were subsequently incorporated into President Kennedy's Executive Order 11007 of February 26, 1962.[32] This order sought to do what the FACA later was to do by statute; its scope and the rules prescribed for committee activity closely resemble the advisory committee act.

Action by the BOB on March 2, 1964, mandated additionally that public advisory committees would be covered by Budget Circular A-63, which dealt with interagency committees. The circular would apply if the committee had "members from two or more Federal agencies in addition to the public members."[33] A further step came on

June 6, 1972, when Executive Order 11671 superseded 11007 in a last ditch effort to forestall congressional action. The new executive order prescribed regulations for all advisory committees, the thrust of which was similar to pending House legislation. Congress, however, was not deterred from enacting the FACA in the fall of that year.

Why and how Congress in the spring and summer of 1972 decided the FACA was necessary remains a minor mystery. After conversing with the very few persons still alive with firsthand knowledge of the actual events, reading transcripts of contemporary congressional hearings, and studying the few retrospective reflections, I am still left with a sense of puzzlement. How did this law actually get passed? And why did President Richard Nixon sign the bill? The answer seems to lie in the change of heart by Senator Lee Metcalf, Democrat of Montana, who switched from leaning against that legislation in the fall of 1971 to being firmly convinced by the spring and summer of 1972 that legislation was necessary.

The episode that crystallized Metcalf's thinking was apparently the dispute over the weed killing chemical, 2,4,5,-T, which contained dioxin. Metcalf and his associates were deeply troubled by the Department of Agriculture's (and the Environmental Protection Agency's) wavering efforts to assess the product's potential danger to the public.[34] The apparently secret initial proceedings of an advisory committee summoned to explore the issue seemed to epitomize in Metcalf's mind the government's secretive approach to a vital public concern. The slightly hysterical press treatment and the general climate of distrust of government then prevailing helped to create a favorable atmosphere for the legislation. Metcalf and his colleagues did not feel that what they were doing was earth shaking. But they did believe they could shake up the executive branch, alter its processes, and put the Congress more effectively into the picture in a watchdog role.

Why the executive branch acquiesced and why President Nixon signed the bill is less clear. Evidently he had bigger fish to fry at the time and did not think that the measure would be taken very seriously. The FACA would not affect very much (just as the Freedom of Information Act a few years earlier had so many exceptions that its impact was limited). William D. Ruckelshaus, as the crusading new EPA chief, apparently also forced Nixon's hand by getting out in front on the issue and embracing the concept of openness in decisionmaking to ingratiate himself with the environmental community.

In any case President Nixon signed the FACA into law on October 7, 1972. Executive order 11686 was thereupon issued, revoking and superseding the earlier executive order, conforming all executive regulations to the provision of the new statute, and designating the OMB as the agency responsible for administering presidential functions under the act.[35]

The FACA overlaps with three other acts: the Freedom of Information Act (FOIA) of 1967, the Federal Privacy Act of 1974, and the Government in the Sunshine Act of 1977. Taken together these acts form the major effort to achieve openness in the federal government.

The FOIA was the first step. It was directed toward documents on the premise that if the public could gain access to the written records of government, citizens could track what agencies were doing or contemplating. The public's confidence in government would presumably be enhanced by knowing the thinking that went into decisions.

The FACA sought to expose the supposedly shadowy penumbra of government to the clear light of day by letting the public know in advance about advisory committee meetings, letting them attend the meetings, and providing a record of actions taken or recommendations made at such meetings. The FACA's exemptions for the closing of meetings paralleled the FOIA's specific exceptions for not releasing documents.

Because the FOIA excluded certain areas of decisionmaking from public inspection, the Federal Privacy Act gave individuals access under specified conditions to their federal records. This act contradicted the assumptions of the FOIA somewhat in that it sought to protect the individual's privacy rights by narrowing rather than widening the amount of information presumed to be in the public domain. By knowing what information the government had about them, citizens could protect themselves from disclosures to creditors, employees, and so on, of information that could be injurious.

The Government in the Sunshine Act, finally, was intended to complete the process by requiring government operations themselves—beyond documents and advisory committee meetings—to be as open as possible. As it turned out, because of the practical difficulties of defining the *meeting* to which citizens might be given access, the Sunshine Act was limited in its reach only to a class of some fifty or so collegial bodies (for example, independent regulatory commissions and multiperson boards) that make decisions by vote at formal meetings.

Key Provisions of the FACA

The FACA, as it was passed in the fall of 1972, was a blend of measures under consideration in the House Committee on Government Operations and in the Senate Committee on Governmental Affairs.[36] The House and Senate versions proceeded from substantially different assumptions. The House champion was Representative John S. Monagan, Democrat of Connecticut, who tended to stress the managerial aspects and the lack of "adequate guidelines, supervision, and direction" of advisory committee operations.[37] Monagan believed the principal need was to assert the government's management control, eliminating the diffusion of authority that he felt advisory committees often represented.

Somewhat inconsistently with the argument that advisory committees were ineffectual and irrelevant, some members of Congress feared that advisory committees had become too powerful and could usurp legitimate functions of government. Advisory groups might potentially constitute "a fifth arm of government" layered on the legislative, executive, judicial, and regulatory or administrative branches.[38]

Senator Metcalf was the most active and influential figure in the Senate side of the debate. Indeed, he remained the legislator most interested in FACA matters until his death in 1978. The FACA was not an issue that engaged the attention of many legislators in either chamber at the time of its passage (or since). Metcalf made the issue a personal crusade.

As the champion of openness, Metcalf had a deep conviction that big interests threatened to dominate governmental policymaking. This conviction shaped the act's provisions for prior notice of meetings, public access to records, and balanced membership. To Metcalf the problem was not preventing the diffusion of government authority but breaking open a closed and unrepresentative policymaking process. Wider public scrutiny and more influence for unrepresented groups were the answers to the dominance of industry in insider politics.

The act ended up with a mixture of House and Senate goals: to strengthen government management of committees deemed to be essential, to eliminate many committees presumed to be of marginal value, and to inform the public about committee membership, activity, and purposes so as to widen the decisionmaking process.

The FACA defines *advisory committee* broadly as any committee or similar group, whose members include one or more individuals who are not full-time federal employees, that is created or "utilized" by the president or by any federal agency for the purpose of giving advice or recommendations. A variety of remedies to prevent abuses is provided, including the requirement for periodic charter and review of all committees, clear delineation of responsibility for the committees' operations, provision for publicly announced and open meetings, and minutes made available to the public.

The act, interestingly, does not contain prohibitions regarding conflicts of interest but deals with the problem by requiring "balanced" committee membership. This would eliminate bias and favoritism because the members would interact and check one another. Government officials establishing advisory committees should "require the membership of the advisory committee to be fairly balanced in terms of the points of view represented and the functions to be performed" so as to "assure that the advice and recommendations of the advisory committee will not be inappropriately influenced by the appointing authority or by any special interest, but will instead be the result of the advisory committee's independent judgment."[39] The act's sponsors believed the deliberative process characterizing the committee's work was the best guarantee against conflicts. Nonetheless, as will later be explained, a thicket of separate conflict-of-interest provisions has grown up and become effectively incorporated into the legal framework surrounding the FACA.

Openness

The FACA's requirement for open meetings is offset by provisions for exceptions in the case of national security, sensitive personnel discussions, and various other criteria. As mentioned, the provisions justifying the closing of meetings were originally the same as those of the FOIA, but in 1977, with the passage of the Government in the Sunshine Act, Congress amended the FACA to narrow somewhat the grounds for closing a meeting. This was done in response to a feeling among public interest groups and their congressional allies that the FACA provisions were too loose and let too many meetings be closed. Some departments also have their own requirements for closing meetings spelled out in their own authorizing legislation.

The Department of Energy has the most stringent openness requirements; it can close meetings only for national security reasons.

As a result, some committee work (for example, the choice of winners of the Fermi and Lawrence awards) is conducted by mail, with the formal meeting simply the occasion for announcing the results.

The FACA requires advanced notice and publication in the *Federal Register* of the times and places of meetings. Usually this has been interpreted by the agencies to mean fifteen days in advance of the meeting, but with printing time and internal agency clearance the notice requirement in practice normally means at least a month. If a committee wishes to meet on short notice, the prior notification can be burdensome. In practice, informal consultations among committee members and the activities of subgroups appear to constitute working accommodations to the notice requirements under the FACA.

Utilized Committees

A troublesome issue in interpreting the FACA has been the application of the law with respect to *utilized* committees. Such a committee is a privately established group that is used by the government as "a preferred source of advice." An outstanding example would be the many advisory functions performed by the National Academy of Science's complex. Members of utilized committees, including all independent contractors who serve the government, are not subject to the constraints affecting advisory committee members.

The utilized committee is subject neither to the open meeting nor to conflict-of-interest requirements. The *representative* advisory committee member is subject to the open meeting but not to conflict-of-interest requirements. All committee members who receive compensation are classed as *special government employees*, or SGEs, and are subject to conflict-of-interest and openness requirements. Because utilized committees are exempt from both the FACA and the conflict-of-interest laws, the status of the utilized committee will likely continue to be a subject of controversy. Critics will try to make utilized committees subject to the same kinds of requirements applicable to other advisory committees. The agencies, along with the utilized organizations like the National Academy of Sciences, Rand, and other think tanks, will continue to resist. Common sense would suggest that research organizations would find it virtually impossible to live with or even imagine what conformity to the FACA would mean for their operations. Such a broad application of the FACA would very likely deprive government of an important dimension of advice; universities and research organizations would think twice about entering into contract arrangements with government. Almost certainly a muddle and a long period of litigation would result. The National Academy of Science's complex has been held to be exempt from the FACA provisions in a decision that significantly narrowed key FACA provisions.[40] Future judicial opinions could, however, widen the boundaries of the law's application.

The FACA has been interpreted to exclude private organizations with which the government merely contracts for advice or studies, informal clusters of individuals who do not constitute a group or who may have come together for only one meeting with a government official, and persons acting informally on their own and in their private capacities. The legislation otherwise takes a very broad definition of what constitutes an advisory committee. Any "board, commission, council, conference, panel, task force, or other similar group," is included, regardless of the number of members or the structure of its organization or composition.

For an act devoted to opening up the governmental process and increasing citizen participation in government, the FACA nevertheless creates a presumption against new committees. An agency head must make a formal determination that a new committee is in the public interest. He or she must "consult" with a central staff agency (initially the Office of Management and Budget and later the General Services Administration), a step which de facto has become a regulatory process. Consultation amounts to the granting of permission even though approval usually has been routinely granted. The central administrative supervision, except for the early Carter period and an interlude in the first Reagan term of office, has been relatively lightly exercised, but many instances exist of requests being turned down. Sometimes this is a case of the OMB or the GSA throwing its weight around, but more typically a denial results from the perception that a particular committee could embarrass the president or be used to lobby for agency appropriations.

Balance

Once determined to be within the meaning of the act, the requirements of sections 5(b)(2) and 5(c) come into play: the committee membership must be "fairly balanced in terms of the points of view represented and the functions to be performed." In an act fraught with controversy, this has been perhaps the most controversial provision. The charge that committees have lacked balance has been the rallying cry of critics on both the political Left and Right. The Left has felt, for example, that advisory committees have too heavily represented industry. In turn other critics have zeroed in on the balance provision as ambiguous, unworkable, and a relic of the leftism of the 1970s. The scientific community has found the balance provision particularly irksome. Must balance be in terms of scientific disciplines, in terms of lay versus specialist members of a committee, in terms of industrial versus academic scientists, or what?

The original intent of Congress seems to have been reasonably flexible; balance was to be defined in terms of geography, that is, a committee should draw on members from different parts of the country. And the words "in terms of the points of view represented" and the "functions to be performed" suggested that a committee might reflect differing professional backgrounds (applied scientists as well as basic scientists, mechanical engineers as well as electrical engineers, academics as well as industrialists, and so forth). The exact mix would depend on the purposes to be accomplished. Congress apparently did not intend balance to be defined in terms of race or gender. Nor did Congress intend that imbalance should be grounds for legal action against the agency creating the committee, although this is less clear. Arguably, if the guidelines for balance are to be taken seriously, a legal remedy in the form of injunctive relief might be implied.[41]

Congress would surely not have intended to bar any purely scientific advisory committee. That scientists as a group lack balance vis-à-vis laymen who represent the public interest is a claim later advanced and litigated with partial success by Public Citizen, the Ralph Nader litigating arm.[42] The Committee on Government Operations of the House of Representatives did believe, however, that more nonexpert, interested, and knowledgeable "public members" and "environmentalists, consumers, geographic representatives, and noninvolved persons" should be included on advisory committees. It also endorsed the requirement of the then-prevailing Executive Order 11007 that "industry advisory committees" be reasonably representative of the industries to which they related.[43]

Whatever the original congressional intent, race and gender became prominent criteria for committee balance beginning in the Carter administration. In the Reagan administration the issue of public interest representation came to be important both for those who feared committees that were composed too heavily of industry members and for critics hostile to the regulatory mission of the agency. A favorite remedy of public interest advocates was to urge that public interest representatives be added to all science advisory committees dealing with regulatory issues. In practice even highly technical advisory committees usually have come to have some lay members. The reason for this is clear: most problems that engage the attention of high-level officials are not exclusively or narrowly technical, and even the aspects that seem heavily technical often have political, economic, and social dimensions as well. Committee organizers know from experience that they need a broad array of skills once a committee's mandate goes beyond a very specific and circumscribed task.[44]

Conflicts of Interest

Among the most vexing concerns is the set of issues relating to the potential conflicts of interest of committee members. A controversy over alleged conflict-of-interest violations of past members of the EPA's FIFRA (Federal Insecticide, Fungicide, and Rodenticide Act) Science Advisory Panel in the summer of 1989 brought a wave of bad publicity, and drew attention once again to the workings of the advisory system. Critics saw the EPA incidents as underscoring the need for major reform of the advisory committee act. A subsequent EPA inspector general report and an inquiry by the Public Ethics Office of the Department of Justice failed to substantiate the charges of conflicts of interest arising from the EPA panel's consideration of the pesticide Alar in 1985.[45] But significant congressional concern has remained; the prospect exists for future legislative action if new incidents come to light.

Several observations are noteworthy at the outset. First, the legal framework is remarkably recent, having taken shape mainly within the past decade. Indeed some of the most dramatic developments occurred only in the late 1980s. The public awareness of corruption in government and the rise of investigative journalism seem likely to guarantee that the spotlight of media attention will remain focused on the "revolving door," "influence peddling," indirect cost abuses, and other real and imagined ills.

Second, the framework of law and regulations relating to science advisers has not resulted from explicit consideration of the advisers' role. Instead the applicable body of law and regulations has largely resulted from efforts to deal with other situations that then spilled over to the advisory situation. The FACA itself, as noted, contains no conflict provisions.

Third, the consequences of the changing legal climate in which advisers have been forced to operate, while not easy to assess, have frequently been negative. In particular, the contrast in treatment of the members of advisory committees that must be classed as special government employees and of those that are considered members of representative advisory bodies is troublesome. The new climate, while seldom preventing the creation of advisory committees, does tend to reduce their effectiveness. This will be true especially when the legal staffs of the agencies adopt a hypercautious approach to applicable law and regulations, or if legal counsel in outside institutions advise their clients against service on advisory bodies.

Legal Framework

The first major effort in the modern period to legislate ethical conduct came in 1962 after an Advisory Panel on Ethics and Conflicts of Interest in Government, appointed by President John Kennedy, reported that ethics laws were uncoordinated and not comprehensive. The panel proposed legislation that was adopted in 1962, the Bribery, Graft, Conflicts of Interest Act.[46] This remains, as amended, the main law regulating the conduct of present and former officials of the executive branch and members of Congress.

The covered government employees are prohibited from "participating personally and substantially" on behalf of the government in any "particular matter" in which they have "a financial interest." A distinction was made between full-time federal employees and "those who serve the government intermittently or for a short period of time" (the category of what came to be known as the special government employee was hereby invented). The legislative history of the act makes clear that Congress was mindful of the contributions of temporary advisers. Congress sought to ensure that the government would continue to make use of outside advisers, including scientific advisers.[47]

In 1965 a General Accounting Office investigation of contractor gifts to defense officials helped to convince the Johnson administra-

tion to issue Executive Order 11222 strengthening ethics requirements.[48] This executive order contained the first requirement for federal
officials to report their financial interests for the record. The disclosures, however, were relatively modest as to both the number of officials covered and the level of disclosure required. Part-time advisers
were covered under section 306 of the executive order.

In 1968 allegations about misconduct by Representative Adam
Clayton Powell led Congress to require disclosures by its own members and top staff.

Between 1969 and 1973 the ethics programs initiated earlier, in the
words of a close student of the evolution of federal ethics requirements, "fell apart because of White House, Civil Service Commission, and agency neglect."[49]

Although the Watergate hearings of 1973 and 1974 did not focus
on conflicts of interest, the concern with ethics raised by the Watergate scandal created a mood broadly receptive to the reform of existing ethics laws.

A 1975–76 GAO investigation of the effectiveness of the existing
financial disclosure system revealed widespread agency neglect.[50] The
GAO findings in turn formed the basis of a Common Cause report
that helped to solidify support for significant new ethics legislation.
The Ethics in Government Act of 1978 was the result of this reform
wave.

This act did not create an extensive body of new law but did take a
number of steps that strengthened ethics requirements. The act wid-
ened the number of officials covered under the conflict-of-interest requirements (some 5,000 officials were now included); made substantially
uniform the disclosure requirements affecting judges, legislators, and
executive officials; required public rather than merely confidential disclosures for certain officials; created the position of special prosecu-
tor—now independent counsel—as well as the Office of Government
Ethics with governmentwide responsibilities (originally located in the
Office of Personnel Management until it was moved to the Executive
Office of the President in September 1988); directed executive agencies to designate an ethics officer to administer conflict-of-interest laws
and regulations; and required the Senate to create an Office of Senate
Legal Counsel.[51] Those filing confidential disclosures were not subject
to the same detailed financial reporting required in the public disclosures. As interpreted by the Department of Justice since 1978, the
ethics laws contain a broad prohibition against advancing one's own

financial interests through actions either as a regular full-time or special government employee, whether the action in question is adopting general rules, rendering advice, awarding grants, or engaging in other governmental activities.[52]

Despite this extensive ethics reform activity, the basic rules applicable to scientific advisory committees were affected only slightly. But in the light of heightened interest in ethics reform, agencies have generally begun to pay closer attention to their conflict-of-interest programs and have administered existing requirements more vigorously and thoroughly. The practical result has been that ethics requirements have appeared to be more stringent, and compliance with the requirements to have become more onerous.

President George Bush, in one of his first official actions, appointed the President's Commission on Ethics Law Reform, which reported on March 10, 1989.[53] On April 12, 1989, President Bush issued an executive order, *Principles of Ethical Conduct for Government Officers and Employees*, that supersedes previous presidential executive orders. This order, in section 503(b), removes the distinction between regular and special government employees for the purposes of the applicability of certain ethical requirements.

The president also prepared and sent to Congress a comprehensive reform measure, the Government-Wide Ethics Act of 1989, in which he sought, following the recommendations of his commission, to broaden the use of waivers to provide protection for advisory committee members from potential criminal liability and thus to maintain government's access to outside technical advice. No action, however, has been taken on the president's proposals. The Office of Government Ethics, with Justice Department concurrence, has issued minor technical regulations.

The congressionally inspired Office of Federal Procurement Policy Act of 1989, together with implementing regulations, forms the final element of the federal ethics law reforms as of late 1991. This statute and its implementing regulations will probably have their most significant effect on the postemployment behavior of regular government employees who served in procurement positions. The aim is to regulate the revolving door phenomenon in defense contracting. Most science advisory committees operate in the very early or upstream stage of government actions and have little to do with procurement decisions. The legislation, however, leaves ambiguous the definition of *procurement* and also the point at which a procurement action begins.

This ambiguity could prove troublesome to officials connected with advisory committees.

Special Government Employees

A vexing aspect of the conflict laws as they apply to science advisers is the different treatment of special government employees and those who are otherwise classified. Advisory committee members are considered special government employees when they meet one or more of the following criteria: they are appointed by the government officials whom they advise rather than nominated by an outside association; they respond to an agenda set by the government; and they receive compensation for their services.

In practice the decisive factor is compensation. Individuals may meet the other criteria and still be classed as representative committee members. As such they are not subject to conflict-of-interest liability or to disclosure requirements. But if an adviser receives compensation, he or she is considered a special government employee even if serving on a representative advisory committee.

The absence of liability for noncompensated members of committees contrasts sharply with the treatment accorded the special government employees. They are subject to potential criminal penalties (principally under 18 U.S.C. sections 202–18, and 5 C.F.R., section 715), to disclosure and reporting requirements, and to a thicket of waiver, recusal, and other requirements and obligations. The paradoxical result is that people serving on advisory committees who clearly have a vested interest in agency decisions do not have to worry about disclosure or conflicts of interest. They have a known biased position, but the government wants to hear their views. Those who are supposed to be disinterested and who clearly do not represent an affected interest must disclose and are subject to severe criminal penalties, including lengthy jail terms, for advising government on matters in which they or members of their family, firm, or institution may be found to have a "particular interest."

This outcome is not as bizarre as it may at first appear. Clearly if the purpose of an advisory committee is to let the agency know what an industry is thinking before framing policies affecting that industry, the aim is not disinterested policy guidance. It is precisely to hear the special concerns of that sector. On the other hand, part of the aim of an advisory system is to offset the parochial and self-serving view-points with disinterested points of view. The science adviser who is supposedly objective should properly be held to higher standards. But these standards should not be so onerous or so erratic in application that people will decline to serve. This would defeat the whole purpose of balancing the interested and the disinterested advice by leaving the self-serving and parochial as the only source of advice.

No person is ever entirely free of potential conflicts, including one who is professionally active in research, consulting, or other noncommercial activities that are nonetheless compensated. For such people to be required to disclose potential conflicts is appropriate (the detail of disclosure and whether the disclosure should be public or confidential are separate issues). However, the maze of current practices, the potential criminal penalties, the constraints on future professional activity, the opportunity for simplistic and harmful publicity, engineered by politicians or others for their own ends, and the plain inconsistencies, uncertainty, and confusion that now prevail with respect to special government employees are counterproductive. The current system is clearly in need of reform. More uniformity in approach among the agencies, or at least within a department, would be useful. (See table 2-1 for a summary of agency practices on conflicts of interest.)

The Evolution and Management of the Advisory System

A committee management regime has emerged to administer the FACA. Endowed with a modest staff, it compiles data and enforces a degree of compliance from the agencies. But whether and to what extent the management system actually manages or even understands the advisory system has been harder to tell. Do the various components of the regulatory regime in the OMB, the GSA, the Office of Government Ethics, Congress, and the agencies provide effective direction or oversight of the vast, sprawling advisory network? Do the government officials who seek the help of science advisers take any notice of the committee management staff units that are supposed to oversee the advisory activity?

The initial implementation of the FACA might be charitably described as loose. The legislation, signed into law on October 6, 1972,

Table 2-1.   **Advisory Committees and Conflict-of-Interest Requirements in Selected Agencies**

Agency	Advisers classed as SGEs[a]	Classed as "representative"	Financial disclosures required	Waivers	Use of recusals
Department of Defense[b]	X		Confidential	Rare	Yes
Environmental Protection Agency	X		Confidential	Some	Yes
National Aeronautics and Space Administration	X		Confidential	No	Yes
Department of State	X		Confidential	No	No
Department of Energy		X	None[c]	No	Yes
Department of Commerce		X	None	No	No
Department of Transportation	X	X	Confidential	Yes	Yes
Department of Health and Human Services	X	X		Yes	Yes
Department of Labor					
National Science Foundation	X	X	Confidential	No	Yes[d]

a. Special government employees.
b. Defense at one point had representative industry advisory panels.
c. But advisers are required to sign a conflict-of-interest statement; the burden is on advisory committee members to be alert to appearances of conflicts. The Energy Department practice is modeled on National Academy of Science procedures. The department requires tighter conflict-of-interest procedures for its full-time employees than most other agencies.
d. Full burden put on individual to avoid conflicts by means of recusal.

was to take effect on January 5, 1973. To say that responding quickly to FACA requirements was not the highest priority on agency minds would be a polite understatement. The initial reaction was sluggish and indifferent.[54]

The president delegated his responsibilities under the act to the director of the OMB (Executive Order 11769), who then delegated his functions under the act and the executive order "in general" to the Committee Management Secretariat in the OMB (OMB Circular A-63). The secretariat was then left on its own with virtually no attention from higher OMB management.

A window into the state of affairs with respect to governmentwide compliance with the FACA is provided by a 1975 report prepared by the second director of the Committee Management Secretariat, Chet Warner. He took office on March 28, 1974, shortly after the agencies were due to send in their first annual reports on FACA committee activity. His initial findings on the status of operations revealed that his office "had received only thirty-one annual reports of the fifty-four due from the agencies no later than February 1, 1974." Further,

"some agencies, when called about the missing reports, had no committee management officers, didn't know about the annual report, and
had never heard of the Federal Advisory Committee Act."[55] As to the
office's own authority, it "did not have a list of delegated responsibilities from the [OMB] Director . . . had no written regulations on its
own procedures or processes . . . had no advisory committee management plan . . . had very few charters or other pertinent information
on advisory committees [and] had two professionals and one part time
secretary to run this operation."[56]

An extensive survey conducted between April and June 1974 was
even more discouraging to Warner:

> To sum up . . . there was very little knowledge about advisory
> committee management in the Congress or the executive branch of
> the Federal Government.
>
> My survey showed [that] the biggest problem in administering
> the Federal Advisory Committee Act was the total lack of interest
> and knowledge on the cabinet and subcabinet level. The policy-
> makers did not know and did not care about complying with the
> Act.[57]

Warner was something of a voice in the wilderness in his own agency.
Even though he managed to bring about considerable improvement in
the committee management system, his frustrations with the OMB's
lack of interest in his work grew over the rest of 1974. His assessment
of the OMB's interest in his work was as follows:

> The best example of total default on the policy making and decision
> making level of the Federal Government regarding the Federal Ad-
> visory Committee Act is my own experience at the Committee
> Management Secretariat, OMB. Not one word of policy on the Act
> was made or written by the Director, the Deputy Director, Assis-
> tant Director, Program Associate Directors, or Deputy Associate
> Directors. . . . The general attitude of the policy makers and deci-
> sion makers at OMB was we don't know anything about advisory
> committees; we don't care anything about advisory committees;
> and you take care of the problems.[58]

Warner was determined not merely to secure full compliance with
the FACA's requirement that agencies file committee charters. He

also sought to test the balance requirement of the act, and for this purpose he found the right ally in Senator Metcalf and the perfect foe in the National Petroleum Council.

Metcalf made a practice of holding oversight hearings in an effort to embarrass agencies he felt were not complying with the act. In oversight hearings in November and December 1973 and again in February 1974, he had pressed the issue of balance and pointed to the National Petroleum Council as a flagrant violator because it had only industry representatives.[59]

Warner, arriving in his post at this time, found in Senator Metcalf the supporter he needed to enforce conformance with the FACA on the executive agencies. He was able to "reason" with the agencies by threatening them with a potential Metcalf investigation or oversight hearings.

This arrangement worked well until the next phase of the Interior-Metcalf fight, which erupted in September 1974 and was to cost Warner his job. The Department of Interior on September 6, 1974, announced the appointment of 21 new members to serve with 113 members reappointed to the council. All were members of the petroleum, chemical, or other allied industries. Most were scientists, engineers, or other specialists. There were no consumer representatives as either new or reappointed members.

Furious, Senator Metcalf wrote to Secretary of the Interior Thurston Morton demanding to know why Interior had failed to honor its commitment to balance the council's membership. Jack W. Carlson, assistant secretary for energy and minerals, replied that the law "does not tell us whether it means balance within the industry or in the society."[60] He noted that the council had always been a technical advisory panel representing different specialists and branches of the industry. Nonetheless he pledged in a speech at the National Petroleum Council's annual meeting that "we will recharter the council to conform to the Federal Advisory Committee Act when the current charter expires at the end of the year."[61]

But Interior was slow to act. Despite prodding from Warner it was not until December 31, 1974, the day the old charter was to expire, that Interior forwarded its new plan to the OMB. Warner did not concur in the charter renewal, insisting that the plan was incomplete, failed to address the balance issue adequately, and was submitted too late to permit careful examination. A flurry of activity ensued that resulted finally in the Interior Department gaining a verbal approval for the charter's renewal from the deputy director of OMB. Over-ruled and rebuffed by his own agency, Warner quietly resigned and left the federal government.

Warner's successor was a career OMB civil servant, William Bonesteel, who brought the system into a mature and smoothly functioning phase. Bonesteel delegated authority to the agencies, established a system of speedy review within the OMB of agency plans, and operated with an insider's knowledge of the limits of his office. He picked no losing fights but occasionally blocked agency efforts to form new advisory panels when he felt that the committee's formation could embarrass the president or work against clearly articulated presidential objectives. The system had perhaps its most trouble-free period from January 1975 to January 1977. It was decentralized, with decisions largely left to the agencies, though there were occasional and decisive interventions when Bonesteel and the secretariat staff judged that presidential prerogatives were at stake.

The next turbulent period came with the advent of the Carter administration. The initial Carter approach to advisory committees brought substantial shocks to the system. The first reform wave was a populist assault, led by the president himself, on the number of advisory committees. President Jimmy Carter and his close advisers from Georgia considered advisory committees to be redundant appendages of the big government they deplored and wished to dismantle. At an initial cabinet meeting the president reportedly asked his cabinet members to "mow down" unnecessary advisory committees so as to reach an overall total of no more than sixty.

Achievement of this goal, unsurprisingly, proved to be impossible. The number of advisory committees was some 1,350 at the start of the Carter administration; the White House alone had 95 committees operating under the FACA when President Carter exhorted his cabinet to strive for 60. Moreover, in the spring and summer of 1977 a sudden growth of committees had occurred as part of the mental health campaign launched by Rosalyn Carter and Peter Bourne, the president's adviser on health issues. Some twenty-five committees had been formed around the country to carry forward the campaign. The issue arose whether these committees should be classed as advisory committees under the FACA and officially chartered.

The overall mental health committee, chaired by Bourne, had been chartered within the FACA (the committee initially was to be chaired by Mrs. Carter, but this ran up against the FACA requirement that federal advisory committee meetings must be convened by a federal official). Given the emphasis on management streamlining and the trimming of unnecessary committees, Carter advisers warned about the public impact of creating twenty-five new White House committees under a project of special interest to the president's wife. A solution was found in the form of an exchange of letters between the White House counsel and the OMB counsel affirming that the committees in question were merely "fact-finding" subunits of the overall committee and therefore were not advisory committees within the meaning of the act.

Subsequently, judicial decisions have affirmed that subgroups of larger committees do not require separate charters, provided that they report to the agency through the parent committee.[62] Under the law subgroups were theoretically to be bound by the same notice, open-meeting, and record-keeping requirements as their parent committees. Agency practices have varied widely and have often been informal and even casual with respect to subcommittees.

The Carter administration at any rate lost its zeal for a crusade against advisory committees. In December 1977 the Committee Management Secretariat was transferred to the General Services Administration (GSA).[63] The move achieved the dual purposes of reducing the head count of the Executive Office of the President and of displacing to the periphery a function that had proved to be frustrating. The administration did succeed in reducing the number of advisory committees from some 1,350 to approximately 1,000, a level that has remained more or less stable since then.[64] Much of this reduction occurred through the consolidation of existing committees (as happened with the NASA Advisory Council, the Energy Research Advisory Board, and some EPA science advisory panels).

Even though the OMB had formally shed the committee management functions, it continued its practice of discouraging the formation of new committees during the remainder of the Carter term, using its budgetary powers either as an implied or an overt threat.

Carter officials also pressed hard for minority members on advisory committees, including scientific and technical committees. The National Institutes of Health's advisory councils and the National Science Board, for example, sharply increased minority and female representation. The number of consumer and public interest advocates also increased on many advisory bodies. The appointing officials in the agencies themselves often came from consumer, public interest, or environmental backgrounds, or were female or minority. They felt comfortable in appointing their friends and colleagues to their committees.

Although not new with the Carter administration, the practice of the White House clearing appointments to advisory bodies for patronage purposes became more fully institutionalized. This practice continued under Presidents Reagan and Bush. Although it shocks some members of the scientific community, to have patronage appointments on science advisory boards is hardly evil. Indeed decisionmakers must work with people they are comfortable with, and high-level contacts will normally enhance a board's influence. But the use of patronage on technical boards does raise the issue of effectiveness, and sharpens the potential conflict between accomplishing a stated objective through the use of advisers and the more diffuse goal of broader citizen involvement in the administrative process.

The early Reagan period was another difficult time for advisory committee operations, not because the administration had any clear goal in this area but because it had virtually no policy. The agencies were left largely on their own, except for having to find places occasionally on the more prestigious advisory boards for people suggested by the White House. As a result some agencies (administrators in the regulatory arena in general and in particular the EPA) replaced what they saw as Carter-era ideologues on advisory boards with ideologues of their own. In the EPA this practice helped to ignite the political firestorm that engulfed the agency and brought the Reagan presidency to its lowest point during its first term (only the Iran-Contra arms sale of 1986 produced a deeper political crisis).[65] The administration was compelled to install more moderate EPA leadership that managed to calm the controversy. In other agencies the advisory mechanisms atrophied, as agencies—undecided on the limits of their own authority to appoint new members—simply ignored the existing committees.

A further set of problems resulted from zigs and zags in the GSA's administration of the FACA. For a time after acquiring the committee management function at the end of 1977, the GSA had little conception of what to do with its new responsibilities. The function was left to languish. In the initial Reagan period the GSA Committee Management Secretariat proved to be autocratic, intrusive, and too detailed in its oversight. The secretariat's leadership made the most of the small empire that the committee management function represented. The secretariat, among other steps, sought to influence the appointment of the agency-level committee management officers, issued detailed and sometimes contradictory instructions, and in the process frequently became embroiled in conflicts with agency officials.

All of this took place below the level of political visibility. It was sufficient to cause confusion but was not of enough salience to attract high-level attention. Only after an effort to oust an agency committee management officer for alleged insubordination backfired did this GSA committee management regime come to an end. A reassignment of personnel within the GSA solved the problem by producing a change in the secretariat's operating style. Calm returned to the agencies' administration of advisory systems.

Meanwhile congressional interest in the administration of advisory committees for the moment waned.[66] Congress continued to show interest from time to time, however, in the potential for conflicts of interest on the part of advisory committee members. Congressional concern was reawakened in 1987 when a Defense advisory committee created to serve the strategic defense initiative (SDI) program failed to obtain a charter as required under the FACA and awarded a series of lucrative contracts to institutions represented on the SDI advisory committee.[67] The Defense Department moved rapidly to reestablish its management responsibilities over the SDI committee. The episode damaged congressional and public confidence in the Pentagon's handling of the conflict issue and had broader consequences as well.

Previously, concern over fraud and ethics issues had mainly focused on the role of full-time defense employees involved in the procurement function and on the postemployment activities of full-time employees. Now congressional concern was extended to the advisory penumbra of the defense agencies. Moreover, the concern spread beyond Defense to other government agencies and drew attention to their advisory systems. Yet an understanding of the unique problems of advisory committees was rare. The advisory system was usually dragged in on the periphery of larger issues. It got the worst of both worlds: the advisory system was tarred with the general suspicions leveled against "influence peddlers," but few had the patience to carefully examine the role of advisory committees.

But the SDI case did bring the effectiveness of the FACA back onto the congressional agenda and in particular stirred interest among environmental and other public interest groups then engaged in litigation with the EPA. Public Citizen took up the issue of FACA reform as a major goal. In 1988 and 1989, Public Citizen worked closely with the staffs of Democratic senators John Glenn of Ohio and Carl Levin of Michigan in drafting a bill, S.444, that would have been the first comprehensive reform of the FACA since its passage in 1972. S.444 proposed major changes in the conflict-of-interest requirements applicable to advisory committees, in the fair balance requirements of the act, in the rights of litigants to bring suit to enforce fair balance requirements, and in the treatment of industry advisory committees "utilized" by the government.[68]

The Senate Committee on Governmental Affairs held hearings on the bill in March 1989.[69] After a year of relatively little activity, the issue reached a peak of interest at a markup session on March 29, 1990, at which Senator Glenn proposed the adoption of an elaborated and extended version of S.444. He cited abuses in Defense advisory committees and in regulatory agencies as reasons for the major changes. Republican members of the committee, notably Senator William Roth of Delaware, objected to the measure. Roth's objections, along with substantial opposition to the proposed changes by the Bush administration and by outside experts, stopped the proposal. The Department of Justice threatened to recommend a presidential veto. Opponents objected to many provisions of S.444, including the proposals to require public interest groups to be represented on all advisory committees and to strengthen the standing to sue for the purpose of enforcing the fair balance requirements of the FACA.

S.444 would also, in effect, have ruled out scientific or technical committees as a class, because they have no lay members. Under the reasoning of S.444 science advisory committees were viewed as inherently unrepresentative. Agency officials objected to these and other provisions as unworkable and as imposing unwarranted burdens on the process of selecting a committee. Since committees are convened for widely differing purposes, including narrow technical purposes, agencies generally should have wide latitude in choosing the members. In general, as will be seen in the case studies, flexibility in committee creation and use is highly desirable. Congressional directives to proscribe or to mandate specified advisory behavior or membership usually work against the best use of advisory mechanisms.

The climate of heightened public awareness in recent years has led to intensified agency vigilance in administering the law. Meetings have been kept open whenever possible. Adequate notice has been given. Has this practice of openness resulted in anodyne and useless discussions or otherwise hampered the effectiveness of advisory operations? There is no simple answer. Some agencies, such as the Defense Department, have been largely exempt from the requirements because of national security, although even Defense must announce meetings in the *Federal Register* and explicitly justify the closing of each advisory committee meeting.

In other cases extensive use of executive session has coexisted with portions of the meetings held in open session (as was the practice with the Reagan White House Science Council and with the current President's Council of Advisers on Science and Technology as well as with many of the agency committees reviewed in this study). The ample opportunity at lunch, in the evenings, or before the formal meeting for informal exchanges of views has apparently had no marked impact on the ability of committee members to engage in frank discussion. In some cases open discussion has apparently had little effect on the candor and vigor of the exchanges. And of course, assembling the committee for a meeting invariably provides built-in occasions when members may speak privately.

Summary

The setting in which science advisers operate has grown increasingly complex. While advisory committees continue to function as an integral part of the governmental process, the obstacles they face have multiplied. The conditions that permit them to perform creatively have been harder to create and to maintain. A more open government should imply that agencies will reach out to citizen groups of all kinds for advice as well as for support. But as advisory groups become important forums for debating and potentially influencing the direction of agency policies, the constituencies served or regulated naturally seek a role in who is appointed to advisory positions, how they relate to the agency, and what advice is given. The policy battles are waged in the advisory penumbra of government as well as in its formal core.

Another broad trend has been the impulse to purge the policy process of special interests and to lay bare to public view the financial and other interests that could corrupt government. A tangle of laws and regulations affecting some (but not all) citizens serving on advisory panels has thus resulted. The system is both cumbersome and uneven in its application, exposing some advisers to serious criminal penalties and not even requiring minimum reporting requirements of others. In light of the growing concern generally with ethics in government and given the presence of the mass media eager to expose wrongdoing, the conflict-of-interest issues seem bound to present continuing and complex legal challenges and administrative burdens for advisers of all kinds.

The requirements of the FACA and of the conflict laws were enacted as part of the clash of values underlying American government. The struggles over reform of the advisory system have continued to be political disputes and not merely debates over neutral principles of management. The role of the science advisers, indeed, ultimately reflects the classic debate over the expert versus the politician. Are the internal affairs of government agencies mainly fact-finding efforts, the analysis of data, and the choice of means to implement the ends agreed upon by the political actors? Or is governmental activity at all levels and in all branches the clash of values reflecting the quest for power to advance competing conceptions of the public good?

The paradox of science advice is that, though there is scarcely any national problem of consequence that lacks some technical component or dimension, virtually none are wholly or even mainly determined by scientific or technical considerations alone. Scientists therefore must aspire to be unique and objective in some sense (or else they have no rationale for participating in the policy process), while at the same time they must conform to the mores and norms of the political process. They must interact with other specialists and with generalist administrators and politicians if they expect to be taken seriously and to have an impact.

The American public appears to cope very comfortably with the paradox of science advice, at times veering toward the "utopian rationality" model proposed by Ezrahi (that is, the notion that scientists are able to identify the scientifically "correct" solutions to problems which are always reducible to the "facts"). But more often the public acts on the basis of the pragmatic rationalist model, based on the notion of scientific knowledge being only one factor among many to be harmonized into a decision. The decision must always make political as well as technical sense, and citizens instinctively understand the limits of technical knowledge, which the scientists themselves sometimes do not.[70]

Decisionmakers, in explaining the rationale of a decision, wrap themselves in the language of analyses, data, information, and science as simple experiential fact because such symbols have broad appeal in our scientific-rational culture. But the public face they put on decisions rarely deceives them into thinking that policymaking in its essentials is anything other than weighing incommensurable and conflicting values, assessing the importance to attach to a multiplicity of factors, and struggling with science as judgment.

The case studies that follow range from success stories to ambiguous advisory efforts with mixed results to outright failure. By showing the system in action, warts and all, my hope is to enrich our understanding and to enhance the ability to make full and effective use of science advisers.

3 ‖ The Effective Use of Science Advisers: The Defense Science Board

MANY CAREFUL OBSERVERS regard the Defense Science Board as the most successful of the science advisory committees. The DSB emerged in the aftermath of World War II after it had become apparent that science would be a vital element in the nation's future defense posture. Initially regarded as a substitute for in-house scientific capability and given little guidance, the board foundered and was in its earliest incarnation disbanded.

The internal technical bureaucracy that replaced it proved to be inadequate, and the board returned to life in a new form. The case study describes how the DSB finally worked its way into a useful partnership with the internal Department of Defense (DOD) bureaucracy, policy-level officials, and the outside technical community. Although it is a success story, it is not a simple tale of bringing information, science, or "the facts" to decisionmakers ignorant of the technical dimensions of complex issues. It is rather a story of how various sets of technical interests, all pushing their agendas and their versions of the technical side of complex issues, collided in an extraordinarily intricate bureaucratic setting. The decisionmakers in this setting were hungry for and were typically saturated with technical assessments of all kinds and struggled to sort out what was and was not relevant to the matter at hand. A multiplicity of actors inside and outside of the formal government attempted to wrap themselves in the aura of scientific authority. Each had some putative claim to special expertise. The decisionmaker's task was to sort out what the problem was, what kind of expertise was needed to cast light on the issue, and whose agendas were concealed beneath the claims to neutral expertise.

The DSB in the end became the potential victim of its own success.

Drawn so closely into the center of policy battles, it began to suffer from the policy gridlocks facing the policy machinery as a whole. The story thus has a paradoxical twist—to be useful, the science advisers must become part of the battle. But as they are drawn in too closely, they may lose their claim to objectivity and to seeing the larger national interest.

Background and Origins

World War II had clearly established the need for scientific advice in the military services. Operations research applied to submarine warfare, convoy scheduling, logistical problems, and other areas had helped lead to the postwar creation of the defense think tanks providing analytical advice and assistance to the individual services.[1]

A need for staff support to the secretary of defense was also evident in planning the overall defense R&D effort, in applying technology to new weapons and missions, and in supplementing and evaluating the technical capacities within the services. The Joint Research and Development Board was established in 1946 to provide this staff support. The board, later renamed the Research and Development Board (RDB) and formally established by statute, was viewed as a successor to the wartime Office of Scientific Research and Development (OSRD). The theory was that the scientific talent which had helped to win the war should not be lost to the nation. And naturally this group of distinguished experts should have the authority to organize themselves as they saw fit.

The arrangement ran into trouble from the start. There were fatal ambiguities in the mission of the board. The board members evidently felt that they had the authority to cut across the existing chains of command, as the wartime OSRD leadership frequently did. Unsurprisingly they ran up against stiff resistance fom the military services. The traditional military planners had their own formula for boxing out the RDB; its work could begin only when "requirements" had been laid down by the military planners.[2]

The board suffered further from confusion over whether its role was to advise primarily on technical issues or on operational issues. Since it was comfortable to do so, the board members initially organized themselves along the lines of their professional subspecialties. They tended to view their role as evaluating the technical merits of R&D proposals in those areas. The operational officials within the military sought a more integrative view on how technologies could contribute to functional needs, such as air defense, antisubmarine warfare, and strategic doctrine. And perhaps more critically, a part-time board of members working only fifteen days a year could not be an adequate substitute for the lack of full-time technical staff competence within government.

To remedy these difficulties and to provide staff advice that was responsive to his immediate needs, the secretary of defense began to rely more heavily on the chairman of the Research and Development Board. The board in the process felt bypassed; both the part-time scientists who were not consulted and the military service representatives attached to the board objected. In 1952 in response to the mounting difficulties, the chairman of the RDB undertook a far-reaching reorganization of the relationships among the board, the secretary, and the military services. He secured a new charter spelling out the chairman's responsibilities more fully and clarifying the jurisdiction of the various advisory subcommittees.[3] For a time the reorganization seemed to improve matters, but the problems reappeared. Finally, a more radical reform was decided upon: abolish the board and create a new office of assistant secretary of defense at the Office of the Secretary of Defense (OSD) level in charge of research. Outside advisory assistance, it appeared, could not substitute for the lack of effective staff operations within the government. The board was accordingly abolished in 1953.

The Foundations of the DSB

The DSB came into existence in 1956 as a new effort to deal with the longstanding controversy over the proper organization of R&D in the Defense Department.[4] The creation of a new board was also a recognition that while an outside advisory committee could not substitute for inadequate internal technical competence, internal staff alone could not fully serve the department's needs for scientific expertise. The second Hoover Commission report, published in 1955, was critical of the conservative approach of the military toward technical competence, and called for radically new approaches in weapons systems and in the organization of military R&D.[5] The Hoover Commission specifically recommended the creation of a committee of "outstanding basic and applied scientists" to advise the assistant secretary of defense for research and engineering on R&D and weapons systems development.[6]

The problems that the Hoover Commission wanted to solve involved responsibilities within the still relatively weak OSD and the lines of authority between the OSD level and the individual military services. An advisory structure could not substitute for staff weaknesses, but neither could the new office by itself serve the secretary of defense's needs. Both an effective internal structure and outside advice were needed. At this time a major interservice rivalry was developing over who would control missile development, and the weaknesses in the OSD structure were becoming increasingly evident. Further steps were going to prove necessary to strengthen the OSD structure, but it was also hoped that the advice of the DSB could be helpful in this context.[7]

Much of the DSB's early effort, as might be expected, was directed toward clarifying its role in the DOD's complex policymaking machinery. The issues were of such importance that they involved substantial White House attention and advisory assistance from the newly established President's Science Advisory Committee. Advisory committees at the level of the individual military services were also important and still influential to a degree, but their roles had begun to wane somewhat. They were believed to have become tainted by interservice rivalries.

A draft charter for the DSB dated June 18, 1956, bearing the signature of Clifford C. Furnas, assistant secretary for research and development, shows the board composed to a large extent of ex officio members of the three service science advisory boards and of the technical advisory panels of the Office of the Assistant Secretary of Defense for Research and Development. Stress was placed on its ability to coordinate other technical advisory panels. It was charged to give "major and immediate attention to the progress and administration of basic research, component research and advancement of the state of the art in areas of interest to the Department of Defense."[8] It was to report to the assistant secretary of defense for R&D.

The board was formally established on December 31, 1956, in Department of Defense Instruction 5128.31 (an organizational meeting had been held in late September before the board's formal creation). The rationale for the board was to provide a coherent and comprehensive approach to DOD decisionmakers on policy options involving advanced technology. The advisory panels connected to the individual military services would continue to exist, but the new board would presumably better serve the needs of the secretary of defense, who sought to transcend interservice rivalries and parochial service viewpoints. The board was to "devote major attention to delineating the scientific opportunities which hold promise of radically outdating present-day concepts of warfare and will exercise the leadership in stimulating and conducting board studies which involve the scientific potential for new opportunities of warfare. Specific advice will be rendered on the Department of Defense research and development program."[9]

Early History

The bureaucratic battles surrounding R&D and acquisition were intense.[10] For several years the DSB was engaged in a struggle to find its place in the complex and evolving machinery for military R&D. The board's major achievement in its early years was simply to survive.

The first crisis came only a few months into the board's existence when it clashed with its client, Frank D. Newbury, assistant secretary of defense for research and engineering.[11] One aspect of the problem was a standard one besetting advisory boards and their patrons: the infant DSB had been set up under the sponsorship of one man (Furnas) who subsequently departed, and another (Newbury) inherited what was to him an alien creature. But the problem was not merely personalities. Furnas's views on the role of science in defense planning were the opposite of the engineer Newbury's. Newbury, a retired Westinghouse production engineer, represented the skeptical, acquisition-oriented approach to military R&D then associated with Secretary of Defense Charles "Engine Charlie" Wilson (who reportedly said, "We don't need basic research in the Defense Department to tell us that the grass is green"). Newbury's experience with scientists inclined him to the view that they were often impractical, ignorant of the downstream aspects of manufacture, and more concerned with the approbation of their peers than their customers (in this instance the military services). He sought to avoid the DSB by ignoring it and not calling meetings, but was finally maneuvered into convening the board. Newbury so provoked the board at a stormy April 4, 1957, meeting that it called off its next scheduled meeting, complaining that its advice was not wanted.[12]

Newbury, for his part, argued that the board was poorly organized and did not provide practical advice. He cited the lack of a chairman with whom he could deal (the board had chosen its own chairman and disputed Newbury's authority to appoint one). The scientists won this battle when the new deputy secretary of defense, Donald Quarles, sided with them. Newbury resigned. The first significant action of the DSB was thus that it managed to get its supervisory official fired.

The resolution of the fight left the DSB with a more difficult task: it had to focus on the complex defense R&D organizational issues, an assignment that was not strictly technical and that proved to be difficult. The DSB finally backed a restructuring plan in which the OSD level would considerably augment its technical capabilities. This was deemed necessary given the range of important emerging issues in the military uses of space, air defense, strategic force structure, and other mixed technical-political matters. A new position, director of defense research and engineering (DDRE), would be created with strong staff capabilities relating to both research and acquisition to advise the secretary of defense and oversee the technical work of the military services. An Advanced Research Projects Agency (ARPA, later renamed DARPA) would be created to fund research projects with potential departmentwide applications that fell outside the specific mission areas of the services. These ideas for structural reforms were picked up and incorporated into the 1958 Defense Reorganization Act. The DSB's first substantial contribution thus was not advice on policy but on organization, a frequent focus of its efforts in coming years as well.

The 1958 act generally provided the statutory underpinning for a greatly strengthened OSD to assist the secretary of defense in policy formulation. Toward the end of the Eisenhower administration, the team of Secretary of Defense Neil McElroy and Director of Defense Research and Engineering Herbert York brought relative stability to R&D policy and organizational issues. But York did not particularly rely on the DSB for advice. His priority lay in making the internal reorganizations work and in addressing the large policy issues involved in strategic force planning and space policy. As the former director of the Lawrence Livermore Laboratory, he brought a vast expertise of his own and broad informal contacts in the technical community which served his needs in the fast-paced post-Sputnik world in defense and space policy.

The DSB survived the transition to Secretary of Defense Robert S.

McNamara in 1961, but only after a review of advisory committee activities by the new secretary had wiped out many useless and inactive committees. In congressional testimony before Senator Henry Jackson's Subcommittee on National Policy Machinery, McNamara stated his intention to get rid of unnecessary committees and announced that he had already abolished more than 400 of them, mostly industry boards left over from World War II.[13] But he was persuaded that the DSB was useful, and its charter was renewed. Harold Brown, McNamara's DDRE, like York a physicist and former director of Lawrence Livermore Laboratory, did not make much use of the board. As Kevin Cunningham concludes, "Despite the prominence of the individual members of the DSB at the time, neither York nor Brown was inclined to pay much attention to the DSB. Not until the third DDRE, Dr. John S. Foster, Jr., appointed in 1965, did the DSB come to life."[14]

Foster took an active role in the board's work, frequently participated in its summer sessions, and, most important, took its recommendations seriously. Beginning in 1967 the board initiated the summer study sessions where members would gather for two weeks to review draft reports and seek to devise solutions to specific problems. By 1969 the DSB had produced sixty-six studies covering a range of subjects. The board was composed of distinguished scientists, engineers, and nonscientists, with some overlap of membership on the President's Science Advisory Committee. Most members were from industry and the academic sector, with a smaller number from federal laboratories.[15]

Maturity

The cental components of the U.S. stategic posture—the triad of land-, sea-, and airborne strategic weapons: the Minuteman land-based missile; missile-launching submarines; a strategic bomber force of some 500 heavy bombers—were in place or planned for deployment by the early 1960s. All were matters on which the DSB had conducted studies and made recommendations. The DSB had by no means determined the evolution of strategic doctrine or force posture, but its studies played a part in devising technical solutions to problems, and its deliberations—particularly after Foster's arrival in 1965—helped to affirm the broad directives of strategic policy. The subjects on which the DSB worked included limited war, chemical warfare, the military role in space, DOD basic research policy, tactical aircraft, expendable jammers, and the encouragement of civil sector technical innovation.[16]

With the advent of the Nixon-era détente, new issues of strategic arms policy arose to engage the board's efforts. The period from 1970 to 1972, under the chairmanship of Gerald F. Tape, was a successful and busy chapter in the board's history. His successor, Solomon J. Buchsbaum (who served from 1973 to 1976), maintained a high level of activity and an influential role for the board. In terms of numbers of studies the DSB reached a peak productivity in 1976 under Foster's successor as DDRE, Malcolm Currie. However, the 1973–76 period was marked by a large number of leadership changes in the Defense Department (from Melvin Laird to Elliot Richardson to James Schlesinger to Donald Rumsfeld) that disrupted the continuity of its access and diminished its influence.[17]

The following period was highly successful, marked by broad prestige for the DSB within the DOD, under Eugene Fubini's chairmanship (1977–79). Fubini and Secretary of Defense Harold Brown had been close friends for some three decades. Brown felt comfortable in relying on the board for a variety of advisory assignments, and by all accounts Brown made more use of it as secretary than he had as DDRE.

The support of the secretary of defense was and is an important factor in the board's effectiveness. But most important are the working relationships between the Office of the DDRE and the board, both the staff relationships and the personal relationship between the DSB chairman and the DDRE. The DSB has obtained its access to the secretary largely through the DDRE, has gotten its work assignments and its staff support from the Office of the DDRE, and has functioned as part of the research and engineering staff machinery of the office of the secretary. In the 1970s the board achieved the most elementary but most critical condition for organizational success: senior officials were actually interested in receiving its advice. And their staff viewed the board neither as a threat nor a distraction and reinforced rather than obstructed the predisposition of their bosses to listen to its recommendations.

During the period of the late 1970s under Fubini's chairmanship, a technique was developed to ensure that any DSB report approved by the secretary of defense carried with it a plan of implementation by the military services. The services were required to file specified reports indicating in detail how they had complied with the DSB recommendations and what further implementing steps were contemplated. Earlier the board had evolved a looser technique of publishing annually in the classified Defense Technical Information Service a list of all recommendations that had been made during the calendar year. This was intended to enable Defense Department officials to review the various recommendations and have a ready checklist and reference at hand. This technique had been deemed necessary as the volume of DSB activity increased. The later innovation of requiring the military services to report on compliance with DSB recommendations was an effort to institutionalize the board's more powerful role. This technique was not universally admired even within the board itself. Some members felt uncomfortable with the notion of advisers demanding action. The practice was gradually abandoned after Fubini left the DSB chairmanship. It was formally abolished in 1986 by DSB Executive Director George Millburn, who served from 1985 to 1987, because he wanted to maintain a clear line of separation between the advisory and decisionmaking processes. Millburn believed that the board would gain in authority if it was seen as being more strictly scientific in its work and less directly tied to operational matters.

Advocates of formal devices to require agencies to respond to advisory recommendations have been impressed by techniques like those used during the Fubini chairmanship of the DSB. Some observers have suggested that agencies and even presidents be required to react formally and officially to distinguished advisory panels.[18] A formal requirement, however, can compel a ritualistic but not a substantive response from the decisionmaker. A device such as that employed by the DSB under Fubini reflects rather than creates power relationships. Once the respective positions of the OSD bureaucracy and the military services shifted, the incentives to implement advisory recommendations shifted as well.

In the 1980s the board's influence waned to some degree as greater authority devolved to the individual services. Department-wide coordinating mechanisms atrophied under Secretary of Defense Caspar Weinberger's more decentralized management style. The recurrent suspicion that the DSB suffered from conflicts of interest also resurfaced in a 1983 DOD inspector general report. It was also the focus of highly critical congressional hearings of the House Committee on Government Operations chaired by Representative Jack Brooks.[19]

In 1986, after a period of rapid expansion in military expenditures under President Reagan, a series of procurement scandals led to the creation of a high-level study commission to review defense management.[20] In the process the relationship of the acquisition function to R&D once again had become a subject of contention. Lines of authority in defense R&D matters became blurred in a fashion somewhat reminiscent of the jurisdictional controversies at the time of the DSB's founding. In a sense the DSB appeared to return to the circumstances that gave it birth; jurisdictional conflicts and clashes called for a clear resolution of organizational roles and responsibilities.

Secretary Weinberger fought the creation of the commission on defense management and only acquiesced under White House pressure. The commission, chaired by David Packard of Hewlett-Packard, a former deputy secretary of defense, diagnosed the problems of the acquisition process as stemming in part from the lack of high-level attention. Among its recommendations was the proposal for a new position, an acquisition "czar" at the under secretary level. This concept derived from the notion popular with Republicans since the Nixon administration that the Defense Department worked best with a politician (for example, Melvin Laird) as secretary and an industrialist (such as Packard) as deputy secretary. The first would handle Congress and relations with the White House and the State Department, the second would manage the department and deal with the defense industry. A science adviser (the DDRE) would provide the necessary technical input to the decisionmaking process. Weinberger fitted the first role, and his first deputy Frank Carlucci was a strong internal manager. His second deputy, the ill-fated Paul Thayer, was a strong manager with a defense industry background. Weinberger's third deputy, William H. Taft IV, who succeeded Thayer when the latter's legal troubles mounted in 1983, was neither a strong internal manager nor a person experienced in dealing with the defense industry.

The technical input was weak as well, for instead of stable leadership at the under secretary of defense for research and engineering level, there was frequent rotation and no clear direction. As in the late 1940s an outside board could not make up for deficiencies in internal technical management. The Packard commission, in proposing an organizational solution to a personnel problem, did not remedy the weak links to the defense contractor community. The creation of the under secretary of defense acquisition (USDA) failed to consider how the new USDA and the under secretary of defense for research and engineering would interact and work together.

Yet Congress eagerly seized on the concept of the USDA as the solution to the procurement problems. When the new USDA and the deputy secretary did not work out a clear division of responsibilities, the reality proved to be far different from what was envisaged. The USDA weakened the deputy secretary's position but did not consolidate clear authority for himself. Weinberger meanwhile resigned and was replaced by national security adviser Frank Carlucci, whose major preoccupation was to adjust force planning to budgetary realities.

While the pace of DSB activity remained high—1987–88 was actually one of four peak periods in terms of the number of reports produced—the board was just completing studies that had been initiated earlier and was finding no audience for the finished product.[21] Internal management problems also began to beset the DSB. Effective supervision of the working groups by the parent board was no longer easy. As a result some efforts of the board fell short of the standards of quality that had previously marked DSB efforts. The DSB has, however, produced such a stream of technical activities that a summary characterization of its work product is difficult.

The DSB in Operation

The Defense Science Board clearly remains one of the federal government's most hardworking and prestigious technical advisory boards. Some two hundred scientists and industrialists have served over the past thirty years. Currently membership is limited to forty-five persons with an additional number of affiliated senior consultants (often advisers who have served previously on the board). Most of the actual working effort is done in task forces. There are some eight to ten major task force reports each year (doubling in the peak years) and sometimes additional, less formal, working documents and analyses. A task force consists of a nucleus of DSB members plus senior consultants and other advisers appropriate to the task at hand.

Members are selected by the secretary of defense in consultation with the DSB chairman, his executive officers, and the senior staff in the OSD. Originally, of the twenty-five members of the board, eighteen were ex officio members who were on other technical boards; in 1957 there were only five at-large members. In the 1980s there have been some shifts in the composition of the board, most significantly an increase in the number of retired military officers and a growth in the number of chief executive officers of companies (and a corresponding decline in the proportion of R&D or engineering officers from companies). Basic and applied scientists, as called for in Furnas's original charter, have constituted approximately half of the membership in recent years.

Membership on the DSB has never overtly been subject to political patronage, though nominations come from many quarters. Some effort is normally made to accommodate requests and find room on some panel or task force for nominees of influential politicians. A unit within the White House personnel office under Chase Untermeyer (and later Constance Horner) in the Bush administration cleared appointments to major advisory boards, a practice that apparently has existed at least since the Carter administration (when efforts were initiated to place more women and minorities on advisory committees).

The caliber of DSB members has remained high, and service on the board continues to accord substantial prestige. Even in the face of recent conflict-of-interest laws and regulations, individuals have rarely declined to serve on the board.

In practice, as noted above, the senior technical official in the Defense Department has the principal role in developing the work assignments for the DSB. But the joint chiefs of staff and more rarely the secretary of defense have also been responsible for work assignments. Requests for DSB studies are normally filtered through the supporting secretariat of the board, the USDA and DDRE, and the deputy secretary before formal approval by the secretary.

The scope and parameters of a study are often the focus of considerable maneuver and jockeying for position. The military services as well as the defense R&D bureaucracy seek to define the parameters in a way that limits the scope of possible outcomes. The services and defense R&D bureaucracy also seek influence during the course of the studies themselves. As a general rule, the players in defense decision-making try to define the scope of a DSB study at the start, to have an input as the study proceeds, and to claim that a study supports their policy positions once it is completed. If the study manifestly contradicts their policy position, they will invoke other studies to refute the objectionable findings or else will argue that the study was flawed. The uses of DSB studies run the gamut from blocking, to delaying, to initiating, to supporting action undertaken by the OSD or other parts of the Defense hierarchy. Cunningham notes that "while a negative review by the DSB is rarely sufficient to kill a program or project, the DSB's criticism is often necessary when the political leadership wants to cut back or eliminate a program."[22]

The intensity of interservice debates can perhaps be inferred from the number of senior (flag-rank) representatives from the military services who attended and presented their views at DSB summer study sessions. The largest number of flag-rank officers participating in DSB summer studies occurred in the late 1970s. At that time it appeared that major reevaluations of strategic programs were being contemplated.[23]

The DSB summer study sessions are often devoted to the review of the most important DSB studies. Individual task groups in these cases scrutinize in detail the findings, general conclusions, and recommendations of the major studies, drawing on expert witnesses and additional outside advisers as appropriate. Substantial revisions may be made, and the report that emerges carries the weight of the full board. On other occasions the summer session may be largely removed from the ongoing work of the DSB and represents virtually independent efforts. The secretary of defense must approve the agenda of the summer study sessions. In all cases, however, the summer session is not merely a planning or brainstorming function such as the Project Jason activities. Instead, scientific talents are mobilized not to think about long-range defense problems or emerging missions but to focus on action-oriented, policy-relevant tasks.

The range of DSB studies reflects the gamut of activities that science advisory committees typically perform: review and evaluation of advanced R&D programs; review of R&D management issues, of the selection and training of technical personnel, and of the quality of DOD laboratories; and broad policy reviews of nuclear or strategic programs.[24]

Current Trends and Problems

The DSB in 1991 was in a state of transition. Aside from the uncertainties resulting from the frequent change of senior officials and from new conflict-of-interest laws, the board faced profound conceptual problems resulting from the end of the cold war. The place of the DSB in the overall scheme of things in the Department of Defense, as well as the nation's future defense needs as it moved toward a "new world order," was unclear.

The DSB only rarely engaged the direct attention of the secretary of defense. Its workload reflected to a large extent the concerns of the senior technical officials, the USDA and the DDRE. But worry over defense R&D evidently prompted the secretary of defense to reorganize and clarify responsibilities in this area. In November 1990 David Abbington, special assistant to Secretary of Defense Richard Cheney, proposed in a memo to Deputy Secretary Donald Atwood the creation of an under secretary of defense for technology, research, and engineering.[25] This new technology czar would assume the powers over research and technology vested in the under secretary of defense for acquisition, John Betti, leaving the latter to concentrate on procurement issues.

Under Secretary Betti, partly in response to the proposed reorganization and partly as a result of the cost overruns that led to the cancellation of the Navy A-12 *stealth* plane, resigned in December 1990. The DOD then announced the intention to nominate DDRE Charles Herzfeld for the new technology czar position which would report directly to the secretary of defense.[26] An internal restructuring of the DSB had already been initiated with the merger of the Defense Manufacturing Board into the DSB earlier in the year.

These steps were an effort to come to grips with the organizational problems caused by the creation of the USDA position following the 1986 Packard Commission report. The underlying problem was diagnosed in an influential task force report by the Carnegie Commission on Science, Technology, and Government as stemming from the earlier management reforms that "reflected a need to strengthen the 'back end' of the weapons acquisition process, including engineering development, manufacturing, contracting, and industrial-base management. But this emphasis on the back end, while necessary and desirable, has weakened the 'front end' of the process, consisting of research, technology generation, and tentative explorations of military applications."[27]

The solution, according to the Carnegie task force chaired by former defense under secretary William J. Perry, was to "reapportion the Defense Department's research, development, test, and evaluation budget to allow for modest but sustained increases (after inflation) in the [basic] and [applied research] . . . at the expense of the . . . detailed engineering of weapons selected for production, even in the face of declining overall RDT&E budgets . . . [and to] establish an entirely different set of procurement procedures for [basic and applied research] contracts from those used for development and procurement contracts."[28]

The relationships among research, development, production, test and evaluation, logistic support, and operations in the DOD have always been complex. The rapid defense buildup of the early 1980s and the budgetary deceleration begun later in that decade have further complicated the situation. The reforms suggested by the Carnegie Commission have begun to address some of the pressing concerns facing the DSB and the defense R&D bureaucracy. Fundamental problems remain, however, as the Defense Department moves from the cold war era and attempts to plan for the future.

The Defense Department has committed itself to achieving a 25 percent budgetary retrenchment, a goal not affected by the Gulf War. Policy reassessments will continue in the wake of the changing definition of the threat that the nation will face. How the research and technology base of the department will be affected is unclear. Some labs will almost certainly be closed or consolidated.[29] R&D might become more urgent in the light of the success of high-tech weapons in the Gulf war. But the roles of the early R&D stages and the later engineering and acquisition stages remain unclear and in dispute.

The role of defense technology in the nation's commercial competitiveness is another issue of great significance, with complex linkages to other policy concerns. The potential contribution of the DSB to clarifying such broad issues is more important than ever to the nation. The kinds of expertise recruited for the board, nevertheless, may need to change if the board is to remain relevant and useful, given the new defense agenda. The board's role and organizational arrangements also seem almost bound to change in an era of changing defense requirements.

As it struggles to redefine its mission, the DSB faces anew some issues it dealt with at its origins in the late 1950s. What is the relevance of technology to defense? What is the most effective blending of technical advice into the policy process?

There are also new and novel issues. One thorny problem is the impact of a growing DSB involvement with Congress on the board's traditional advisory relationship with the secretary of defense. Board members know that their principal route to influence is through their executive client and patron. They do not wish to have talk-show-type influence or to be kibitzers. The real coinage of their power is the ability to influence decisions. But they have found themselves over the past decade becoming more deeply involved in responding to congressional requests and in having their work embroiled in public debate. This trend culminated in the directive attached to the fiscal year 1990 military construction authorization bill that barred funds for the B-2 pending the submission to Congress of a DSB report. The report would certify that the bomber had met the stealth requirements promised by the Department of Defense.[30] This congressional assertion of power put the DSB squarely in the middle of a tug of war between a Republican administration and a Democratic Congress over an issue of deep concern to both sides.

The background of the DSB's involvement with Congress is worth brief explanation because the cross pressures the board has faced illustrate a central dilemma of the science adviser. The board first became involved with Congress when chairman Eugene Fubini in the Carter administration testified routinely on a number of occasions to raise the board's profile and to foster a wider understanding of its role. The fact that the board had gained an influential role with Secretary of Defense Brown made it an object of growing congressional interest. Secretary Weinberger expanded the congressional ties by inviting key congressional staffers to summer session briefings on important board studies. He did so to mobilize support for his policies and to consolidate ties with the Republican leadership in the Senate.

The relationship took a new turn in 1985 when a DSB member, John Deutch, maneuvered to have Congress request a study of the Midgetman missile in an act of policy entrepreneurship designed to save the missile from cancellation.[31] The Midgetman had bipartisan support, so that Deutch, an MIT provost and a Democratic appointee in the Carter administration, could carry out the study without being accused of instigating a partisan attack on Secretary Weinberger's policies. The study found enough merit in the small missile program to prevent its cancellation. Whether the outcome is seen as good for the country will depend among other things on how important one thought the Midgetman was for national security at the time. But few observers can feel comfortable with the idea of a science advisory body serving as an instrument to block executive action and add further delay to an already excessively pluralist policy process.

The Republican loss of the Senate in the 1986 congressional elections sharpened the partisan disputes between Congress and the administration. Congressional efforts to invoke the authority of the DSB to discredit the administration's position on the *brilliant pebbles* strategic defense initiative proposal took the board's involvement with the Armed Services Committees a step further. The results of the *SDI Milestone* studies of 1988, led by board chairman Robert Everett, were leaked and made their way into the public debate. Some of the conclusions appeared to support the criticisms Senator Sam Nunn of Georgia and others had made of administration policies, or so the critics contended. The dispute could not be resolved clearly because the studies were classified, because official Washington does not have the attention span to digest large amounts of technical information, and because the issues were sufficiently complex as to make unambiguous conclusions very difficult. But as Cunningham discusses, the *SDI Milestone* studies touched off a minor furor when Senator Nunn wrote Secretary of Defense Weinberger complaining about efforts to pressure the DSB members. Weinberger shot back that "it would be most unfortunate if our advisory groups were inhibited in their work by a concern about their interim drafts appearing in public."[32]

Advisers and Policymakers

The dispute over the B-2 bomber, like the SDI issue, had no clear resolution. But there is a moral. These events point to the central issues of the adviser-policymaker interaction. For the advisers the questions are the following. How can we keep our close relationship of trust and confidence with our sponsor if we are drawn into the orbit of congressional or media attention? Is an issue so important to the nation that we must speak out even at the cost of losing our influence and our position as inside advisers? Can we have it both ways by working with both Congress and the administration and attempting to be honest brokers?

For the sponsoring officials the issues are reversed. Can we trust them? Are they maneuvering to nudge the debate along lines congenial to our political enemies while pretending to objectivity? If so, how can we block their efforts without appearing to suppress important information?

It can be taken for granted that decisionmakers will find a way to negate efforts to force them to accept unwanted advice. All the political actors know this and usually act shrewdly to protect their positions. Though nearly all those involved in this intricate advisory minuet understand its nuances and complexities, in public they hew to the line that decisions flow directly from the presentation of technical information and not from politics and partisanship.

For the public, the larger question may be stated as follows: is American democracy well served by the pretense that complex issues can be resolved if only enough scientific information is presented to policymakers? (It is one's own side of course that always portrays the issue "objectively" while the opposition is invariably partisan or political.)

In my view, the democratic dialogue is not well served by the simplistic notion that the advisory relationship is based just on objective information. Biases are almost unavoidable in the selection of facts, the ordering and interpretation of evidence, and the assessment of the significance to be drawn from analysis, inquiries, or experimental tests. Controversies with policy implications can rarely be resolved by referring the disputes to supposedly neutral experts. Quarreling experts can in fact prolong and exacerbate controversies to the detriment of political compromise and consensus. As James Allen Smith observes, "The experts, far from limiting debate and innovation, have created an environment in which so many arguments contend that no consensus is possible. Their never-ending controversies leave even closely attentive citizens in despair of ever coming to agreement on the most important issues."[33]

The public understands more than the politicians and the pundits believe. In their more lucid and honest moments the politicians know that the facts rarely speak for themselves. Yet they cling to the notion of some truth outside the battle to help moderate the intensity of the political struggle. In this sense the appeals to the authority of science at least anchor the partisan warfare within some framework of reality and facilitate political compromise.

The deep conceptual and philosophical problems blend into a whole set of practical concerns. Among the practical problems that the Defense Science Board faces in the near term is the persistent criticism that its membership is too narrowly based. Although the board struggles for balance and for a representative membership, a close analysis reveals that the members heavily represent defense industry and former DOD officials, including retired flag-rank officers. The weight of defense industry representation on the board has grown more pronounced over the past decade. But should it be increasing, especially in a time when the defense establishment will be downsized?

The logic behind the DSB's creation was to have an outside, unbiased, and fresh perspective on a range of important issues. The intention was not to replicate the kind of advice flowing to senior decisionmakers from the military services and from the defense industy. But the board now is like a corporate board made up only of members of its own industry and thus unable to bring a larger perspective on company problems. Knowledgeable observers argue that the board's recruitment practices produce individuals with close familiarity with complex defense issues. This argument has some weight, but the board needs more young, fresh, and disinterested talent.

A related problem is the persistence of conflict-of-interest stigmata, though not in the narrow sense of members seeking to advance their own or their firms' interests. The possibilities of board members advancing the interests of specific firms in their deliberations range from remote to nonexistent. The problem has a more subtle dimension: board members tend to have a mindset that approaches problems from a "big hardware" or "technology fix" point of view. As a result, the board does not imaginatively reflect interests broader than the traditional concerns of the defense industry. The increase in the number of defense industry chief executive officers serving as board members in the 1980s has exacerbated the general problem. In the 1960s the problems of this sort were handled by having two boards, an Industry Advisory Board in addition and in parallel to the Defense Science Board.

The blend of resources, internal and external, needed to provide for an effective science advisory system in the DOD requires further careful thought. The effectiveness of the office of the DDRE as a departmentwide technical agency, the role of DARPA as a leading-edge technology support mechanism, and the relation of these two offices to each other and to the acquisition function present thorny conceptual and practical problems; the roles of the proposed under secretary for research and the deputy secretary as well as the evolving roles of the individual services also pose problems.

As one of the most effective advisory bodies in the federal government, the DSB will probably find a solid niche. Beyond the short-run issues of structure and institutional relations the defense establishment faces potentially far-reaching changes in mission and doctrine. In this context the need for imaginative, long-range thinking—for advice that synthesizes policy and technology and is not narrowly caught up in the struggle for bureaucratic survival—is greater than ever. The DSB can continue to serve important national needs, provided that it can enlarge its vision and adapt to the swiftly changing defense environment.

The kind of changes required are especially difficult for an internal bureaucracy to formulate and implement on its own. The board has shown in its Midgetman study, in its report on the national aerospace plan, and in its *SDI Milestone* studies that it can criticize programs favored by the president and the administration. At its best the science advisory board can help to reformulate policy and evaluate existing programs, qualities likely to be in great need as the Defense Department struggles to define its new mission in the postcommunist era.

4 ||| Science, Law, and Regulation: The EPA Science Advisory Board

SCIENCE ADVISERS HAVE BECOME, in Sheila Jasanoff's phrase, the "fifth branch" of government in regulatory policy.[1] Moreover, few regulatory agencies have such critical need for science advice as those dealing with the environment. Virtually all the issues these agencies are concerned with have a large scientific component and involve assessments of risk to human health and to the environment that seem on the surface to be beyond the ken of laymen. The Environmental Protection Agency will be discussed here to illustrate the complexities of science advice in a regulatory agency, and the EPA Science Advisory Board (SAB) will be the focus.

Despite the special complexities, the issues raised in the context resemble those just discussed in the case of the Defense Science Board. Should the advisers stick closely to science and eschew politics? How independent should the science advisers be from normal operations? Is the science adviser more effective when operating at arm's length or when deeply involved with the agency?

The SAB went through an evolution that in many ways closely resembles the DSB's development. When the SAB started out, scientists defined their role as advising the agency on how to support science. The agency found this posture self-serving and of little use to its mission. Gradually the board became more useful as it learned the intricacies of EPA's mission.

The clash between science and politics is also evident. The board was at its low point early in the Reagan administration when it was suspected of harboring leftist, Carter-era environmental activists. Its fortunes revived thanks in part to the agency's public adherence to the scientific utopian myth of pure rationality applied to regulatory policy, combined with the quiet development of political rapport be-

tween the board and the agency leadership. In the end the board became a useful body to the agency not through its "pure" scientific
credentials but by developing expertise in *regulatory science*, a unique
mixture of expertise and policy relevance. The work of the board did
not satisfy academic specialists who insist on the utmost scientific ri-
gor, but it became useful to the agency. At the same time the scientific
aura still insulated the agency from critics who deplored what they
claimed was the agency's past disregard of evidence in an effort to
accommodate the administration's political line. The SAB's most influential role was ultimately to help reorient the agency's planning and
priority setting.

Science Advice at the EPA

The Science Advisory Board is one of the thirteen standing advisory committees that serve the EPA, along with a small number of
temporary, negotiated-rulemaking advisory committees.[2] All the
committees are involved with the technical issues of regulation, with
R&D activities, or with the implementation of other technical programs. The SAB itself is composed of an executive committee, eight
standing committees, and a host of subcommittees (both standing and
ad hoc).[3] In 1989 the SAB held fifty-four full meetings requiring 612
travel authorizations, also held many teleconferences, and issued thirty-
eight formal reports.[4]

Since the late 1980s the SAB has had some sixty members, principally scientists and engineers from a variety of disciplines, serving on
its various committees. The chairmen of the standing committees
compose the SAB's executive committee. A roster of more than two
hundred consultants supplements the direct membership. The SAB
was created originally by administrative order in 1973. Since 1978 the
board has operated under the statutory mandate of the Environmental
Research, Development, and Demonstration Authorization Act. It was
originally set up to consolidate the fragmented advisory structure inherited from the component bureaus that formed the EPA and has
sought to function as the agency's lead science advisory unit. The SAB's
analysis of itself, conducted in 1989, estimated that 50 percent of the
EPA's major activities in one form or another are debated, reviewed,
or influenced by the SAB.

Ambient air quality standards, for example, must be reviewed by
the SAB's Clean Air Science Advisory Committee (mandated under

the 1977 Clean Air Act amendments). The Clean Air Committee "shall provide independent advice on the scientific and technical aspects of issues related to the criteria for air quality standards, research related to air quality, sources of air pollution, and the strategies to attain and maintain air quality standards and to prevent significant deterioration of air quality."[5] The SAB reviews the scientific basis of as many other major environmental standards as the EPA administrator requests and its work load permits.

As stated in its charter, the SAB's responsibilities include:

reviewing and advising on the adequacy and scientific basis of any proposed criteria document, standard, limitation, or regulation under the Clean Air Act, the Federal Water Pollution Control Act, the Resource Conservation and Recovery Act of 1976, the Noise Control Act, the Toxic Substances Control Act, the Safe Drinking Water Act, the Comprehensive Environmental Response, Compensation, and Liability Act, or any other authority of the Administrator.[6]

To understand the SAB's role it is useful to contrast it with the more narrow science court role played by the FIFRA panel and the informal, negotiated-rulemaking model used by the EPA in other cases. The SAB incorporates elements of both approaches but has a broader role that sets it apart.

The FIFRA Science Advisory Panel as a Science Court

The FIFRA Science Advisory Panel, created under the 1976 Federal Insecticide, Fungicide, and Rodenticide Act, has operated independently from the SAB and its constituent committees. But efforts under way by EPA administrator William Reilly have sought to produce closer working relations between the FIFRA panel and the SAB.[7]

The differences between the FIFRA panel's and the SAB's style of operation illustrate different approaches to the scientist's role in environmental regulation. The FIFRA panel, driven by its legislative mandates and its origins in the politics of pesticide controversies, has chosen to operate more like a policy-neutral science court. That is, it has addressed narrowly the scientific adequacy of the proposed agency actions under section 6(b) (relating to cancellations and changes in classification of pesticides and their potential removal from the market) and section 25(a) of FIFRA (proposed and final regulations concerning pesticides). Further, it has usually attempted to operate on an arm's length relationship with the agency.

According to the legislation, the FIFRA panel's comments must be considered by the EPA, along with comments by the secretary of agriculture, before final action is taken regarding pesticide regulation. The FIFRA panel takes very seriously the legislative requirement that its advice be independent. It insists on reviewing agency actions only after a certain degree of "ripeness" has been reached; that is, after a proposed rule is presented or action of some formality has been taken. The panel has generally felt that its independence would be compromised if it provided advance informal guidance to the agency on test criteria, scientific methodologies, or risk assessment techniques. Since the panel is limited by statute to seven members, it also has had to economize on its time by eschewing any kind of advisory opinions.

A glance at the FIFRA panel's schedule for 1988 and 1989 (it held five meetings in each of these years) illustrates the character of its work. The major efforts were section 6(b) actions (scientific issues arising in connection with agency reviews of the pesticides Aldicarb and Carbofuron); proposed revision of guidelines on testing procedures on the immunotoxicity testing of biochemical pest control agents, on neurotoxicity, and on mutagenicity; scientific issues in connection with a proposed tolerance assessment system for evaluating acute dietary exposure to a pesticide using Aldicarb; scientific issues on pesticides on the oncogenicity classifications of Acetochlor, Dichlorvos (DDVP), Simazine, Express, Permethrin, Cinch, Paraquat, Atrazine, Isoxaben, Prochloraz, Rotenone, Bifenthrin, Clofentezine, Haloxyfop Methyl, and Propiconazole; and scientific issues in connection with a proposed rule on experimental use permits dealing with biotechnicals.[8]

The panel has served a useful purpose in enhancing the quality of internal EPA reviews and in bolstering the agency's public image as a scientifically credible regulator. In a large majority of cases, the panel has supported the agency's internal risk assessments and proposed actions. In controversial cases it has done less well and has not been able to sustain public or congressional confidence in its objectivity. In the few highly controversial cases when it has criticized the agency's internal scientific assessments, the panel has gone beyond its narrow science court role. An example of this was the continuing dispute over the herbicide 2,4,5,-T in the late 1970s (which led to Congress amending the FIFRA to give new focus to the panel's role).[9] The 1985–89 Alar dispute was a further illustration of a protracted dispute over the scientific basis of proposed agency action.[10]

The difficulty with the science court mode of operation is that with intense controversies it may substantially delay agency action without resolving the underlying issues. The Alar case is a classic instance. In 1985 the FIFRA panel found the agency's proposed ban on Alar (a preservative chemical and ripening agent used on apples) to be based on inadequate scientific evidence. The EPA then spent four years on new animal tests and scientific evaluations before affirming its initial finding. Alar meanwhile became caught up in an intense controversy stemming from a National Resources Defense Council campaign and a 60 Minutes television show. The NRDC effort finally led the chemical's manufacturer, Uniroyal Chemical, to withdraw Alar from the market in the summer of 1989 before the panel reached its conclusions. The panel's concern with methodology and with "good science" simply was not credible to the contending parties or to the attentive public, and its deliberations were not timely.

Alar epitomizes the fractious nature and intensity of pesticide disputes. Rules are slow in coming, and when they finally are issued the contending parties refuse to accept the outcome and go to court. Challenges to EPA rules in this area may come from either the environmental interest groups that consider the standards too lax or the affected industry that views the standards as too severe.[11]

The science court mode of operation, in short, has severe drawbacks. In practical terms there is little or no chance that a science court could definitively settle the issues in cases of intense controversy, and certainly not in a timely fashion.

Negotiated Rulemaking Advisory Committees

An alternative approach that the EPA has employed represents an effort to bypass some of the complexities of scientific policymaking by directly engaging clashing interest groups in the policy process. Partly in response to the problems of delay and endless litigation, beginning in the late 1980s the EPA has experimented with negotiated-rulemaking advisory committees. Under this concept, the agency in effect delegates to the affected interests a share of its rulemaking authority in the hope that they can agree preliminarily on a proposed rule. If the parties can reach tentative agreement, the agency, while reserving the right to act or not to act on any recommendations, then uses its authority to issue a formal rule (specifically, a notice of proposed rulemaking published in the *Federal Register*).

The agency is less concerned in this case with the adequacy of the scientific underpinning of the rule than with the attitude of the parties in dispute. If the parties can agree on a proposed course of action, the chances of litigation after the rule's issuance are reduced. Disputes over "good science" are sidestepped under the negotiated rule concept. The larger aim is to reach a timely and authoritative regulatory outcome even if the scientific issues are not all fully resolved.

An example in 1990 was the Volatile Organic Chemical Equipment Leak Rule Negotiated Rulemaking Advisory Committee. Some twenty-five members appointed by the EPA deputy administrator represented the chemical manufacturing industry, the petroleum refining industry, environmental and public interest groups, state and local air pollution control agencies, and the manufacturers of pumps and valves. An EPA official served as convener and facilitator. The advisory committee generally operated in accordance with the open meeting and public notice requirements of the Federal Advisory Committee Act. No effort was made, however, to control informal subgroup meetings of members of the committee. As a representative advisory committee, it was also exempt from conflict-of-interest requirements.

It remains to be seen how far the concept of negotiated rulemaking can be extended. Indeed, it is not certain that the concept can survive legal challenge on the grounds of unconstitutional delegation of powers, evasion of FACA limits, or conflict-of-interest violations. But the early experience has been favorable. The negotiated-rulemaking advisory committees have seemed a useful addition to the advisory system; they offer promise of achieving faster agency action with less postaction litigation.

Other scientific peer review activity has employed similar concepts of face-to-face confrontation and argument among advocacy groups, with only light-handed agency guidance of the proceedings. The process used by the Department of Energy to reach agreement on the laser isotope-separation approach to enriched uranium fuel production is such an example. In this case the DOE, after peer review deliberations lasting eight months and involving government, industry, and the national laboratories, chose the laser isotope-separation method advanced by the Lawrence Livermore Laboratory over its own gaseous diffusion concept and over alternative industrial approaches as the nation's preferred approach to uranium fuel production.

The EPA Science Advisory Board

The SAB can be viewed as performing functions that include the FIFRA panel's science court role and the problem-solving functions of the negotiated-rulemaking committees. It has played an increasingly important role in helping the agency—and in particular the last Reagan EPA administrator, Lee Thomas, and President Bush's William Reilly—to reshape the EPA's priorities. This has been perhaps its most important contribution to helping refashion the agency's way of doing business. In this respect the SAB has gone well beyond the narrow function of commenting on the technical adequacy of criteria documents and proposed rules and beyond the function of proposing solutions to specific operating problems. It has helped produce a change in the EPA's organizational culture.

The SAB's role should not be exaggerated. It would be too much to say that the SAB has come to operate with full clarity of purpose or with a shared sense of mission among all of its members and with senior agency officials. The board did not play the principal role in changing the agency; it did not even initiate the major changes. Its role has been instead to support the policy shifts sought by administrators Thomas and Reilly. Nonetheless its role has been significant. In recent years the SAB has finally begun to perform the role envisaged for it in the years after its creation in 1973.

These developments have taken place without the board's abandoning the core function of gatekeeper for the scientific quality of agency operations. The SAB tried to function as an "apolitical elite,"[12] originally eschewing politics and policy preference of its own. But almost from the start board members recognized that considerations beyond science itself were relevant. Indeed, if the SAB was to be useful, it would have to weigh economic and cost factors, among others, in assessing such concerns as the "health effects" on the most "sensitive groups in the population" so as to ensure an adequate "margin of safety."[13]

The tension has remained between the need to preserve its own and the agency's scientific credibility by ostensibly avoiding policy, and the reality of having to be "in the battle rather than above it" in order to be genuinely useful.[14] The board's 1989 self-study struggles with the issue. The SAB, the study concludes, "operates on the conviction that its mandate is to assist the EPA in marshaling the very best science available for protection of the environment. . . . [T]he Board's obligation is to focus on scientific and technical aspects of issues, and refrain from making public policy judgments—it is the *Science* Advisory Board."[15]

Further the SAB's executive committee should watch to see if "a committee has strayed into policy territory."[16] Yet because the board is a body of "human beings evaluating uncertain science in the very real world, the science-policy separation cannot be absolute. . . . [T]he technical aspects of many implementation and communication issues are hybridized with political, or managerial, or funding, or other non-scientific aspects in ways that create difficulties for SAB review. . . . [But] there is potential for backlash if the SAB is publicly perceived to be meddling in extra-scientific policy and managerial matters."[17]

The self-study recommends that "the SAB should explore these domains vigorously but cautiously. The question is not whether the Board could generate answers (naturally it could); the question is whether by doing so it might stretch and weaken its natural-science authority."[18]

This handwringing has not prevented the board from moving in practice toward helping the EPA to think strategically and set overall priorities. The board in attempting to achieve more coherence in its own activities through a larger role for its executive committee has mirrored the EPA's struggles to overcome entrenched professional fiefdoms. The board's struggles to achieve a larger strategic vision replicate the agency's drive for self-definition.

The Background

Since the story of the EPA's creation by reorganization plan on December 3, 1970, after a period of intense debate, has been told elsewhere, a detailed review is unnecessary.[19] But it should be recalled that President Nixon decided against a reorganization of the Department of the Interior to form the lead environmental agency. Instead he chose to accept the advice of a working group of the Ash Council (the presidential advisory commission headed by the industrialist Roy Ash) and to create an independent agency that would coordinate the administration's environmental initiatives.

The new agency was made up of functions transferred via the reorganization plan from a number of existing departments. The agency thus had no organic statute setting forth its mission. It instead absorbed a variety of technical programs, each with its own mission and clienteles: the pesticide program from the Department of Agriculture; the National Air Pollution Control Administration from the Department of Health, Education, and Welfare; the water pollution functions of the Water Quality Administration from Interior; the interagency Federal Radiation Council; and others. On July 6, 1970, Nixon sent Reorganization Plan No. 3 to Congress, only slightly modifying the Ash council proposal. On December 3, 1970, the Environmental Protection Agency was formally established.

Many of these programs were oriented toward research objectives, reflecting the faith in basic science of the early postwar era.[20] Clearly both integrative organizational arrangements and a new style of agency operations were needed. A thirty-year-old lawyer, assistant attorney general William D. Ruckelshaus, who had the strong backing of his boss John Mitchell and who had had some environmental enforcement experience in Indiana, was chosen as the EPA administrator. Ruckelshaus wanted to hit the ground running and sought ideas on how to organize his new agency.

To address the problem of fragmentation, Alain Enthoven, a former Defense Department official and White House consultant, prepared an organization plan for the new agency that emphasized a functional division of responsibilities.[21] Douglas M. Costle, the Ash Council staff member in charge of the environment and later President Carter's EPA administrator, modified Enthoven's plan on the theory that a change of this magnitude was too much to accomplish in one bold stroke. Confusion and chaos could result. He proposed instead a three-stage approach: in phase one the program areas transferred from the various agencies would be left intact; in a second phase new functional divisions would be added; and finally in phase three the old offices would be abolished and merged into the functional units.[22]

Ruckelshaus opted for the Costle approach. Initially he created five commissioners, each one heading a program area. In April 1971 he undertook an extensive reorganization, closely paralleling Costle's phase two, in which the agency was divided into five divisions. Three were organized along functional lines (Enforcement and General Counsel, Research and Monitoring, Planning and Management), and two divisions reflected a partial consolidation of the old program areas (the Air and Water Division, and the Pesticides, Radiation, and Solid Waste Division).[23] He left agency reorganization in that state. But paying no further attention to rationalizing the agency's structure had long-term costs. The fragmentation of EPA's approach to the environment, its clashing professional perspectives, and its lack of a strategic sense became ingrained problems that continued to plague the agency for years.

Nor did Ruckelshaus devote major attention to the quality of the scientific laboratories he inherited or to the underdeveloped state of pollution control technologies. For an agency heavily dependent on science, it would have been logical to emphasize research and development as an initial goal. Ruckelshaus's background as a lawyer, his law enforcement experience at both state and federal levels, and his assessment of the political realities of his position led him to emphasize enforcement as his top priority. He reasoned that the best way to establish the agency's and his own reputation—and to mollify Congressional and environmental critics of the administration—was to press ahead rapidly with enforcement actions.

In his first sixty days at the EPA he initiated five times as many suits against polluters as the agencies he inherited had done in any similar period.[24] The preponderance of enforcement actions was in the water pollution area where it was easier to sue than under the 1970 clean air amendments. Ruckelshaus achieved some dramatic successes and made a notable public impact for EPA. But as a negative side effect of this early success he also set in motion an adversarial relationship with the White House staff and the Office of Management and Budget.

In May 1971 OMB director George P. Shultz instituted the "quality of life" review procedure. This was the first of a series of procedures under which the OMB would review proposed EPA and other agencies' regulatory actions to ensure that they conformed to presidential policy. Economic development and fiscal concerns were goals that the OMB sought to have reflected in agency decisionmaking. A similar review process under President Carter was conducted by a body called the Regulatory Analysis Review Group, and by the Regulatory Reform Task Force under President Reagan.[25]

This process undoubtedly delayed the issuance of environmental regulations and from time to time triggered acrimonious disputes between the White House and Congress. The proper scope of OMB powers became a point of serious friction. Congress attempted to restrict the funding for the OMB review office but eventually acquiesced in the assertion of power over rulemaking.[26]

Ruckelshaus in any event could not escape the issue of EPA's scientific credibility. Even as he sought to capture public attention through vigorous enforcement activity, disputes over the health effects of lead and especially over the herbicide 2,4,5,-T containing dioxin confronted him early in his tenure.[27] The conflicts illustrate the close connection between the agency's overall reputation and its scientific credibility. The disputes also illustrate the difficulties of establishing scientific credibility and the limitations of the concept of good science as an agency goal.

Every EPA administrator, as well as other regulators, has embraced the concept of good science at least as a rhetorical goal (as indeed have nearly all outside critics). For most political actors, however, good science means the desire to maintain the appearance of scientific integrity while scoring points against the supposedly biased opposition. Politics has usually overwhelmed any quest for objectivity or the desire to educate the public about risks. Good science simply cannot resolve all the uncertainties that exist in regulatory issues.

The 2,4,5,-T Controversy

On April 15, 1970, the secretary of agriculture announced a limited ban on the use near the home of the herbicide 2,4,5,-T, for health reasons, but did not limit its use in agricultural applications. This action came in response to studies by American scientists critical of the Vietnam War that focused on South Vietnamese children exposed to defoliants using dioxin, as well as to a National Cancer Institute study released in 1969 that was based on laboratory animal tests. The studies by both groups found some possible links between birth defects and 2,4,5,-T.

Industry and environmental groups both objected to the secretary's decision, the environmentalists because the ban was too limited, and industry because there was a ban at all. Scientists lined up on both sides of the issue, which recalled the disputes over nuclear weapons policy and over the risks of civilian nuclear power that arrayed scientists against one another.[28] The surgeon general, the president's science adviser, and the Herbicide Assessment Panel of the American

Association for the Advancement of Science were among those supporting the partial ban. On the other side, Emil Mrak, chancellor of the University of California at Davis and a pesticide expert, argued that "2,4,5,-T had been the victim of a panic-button operation," and the Weed Science Society of America denounced the ban as "sensationalist." Two manufacturers of the herbicide, Dow Chemical and Hercules, Inc., called for the creation of a scientific advisory committee to review the evidence.

In an effort to quell the controversy, a panel of scientists under the chairmanship of Mrak was convened to review the evidence on which the 2,4,5,-T decision was based. The panel completed its work in May 1971. With pesticide regulation having been meanwhile shifted to the EPA, the panel transmitted the report to Ruckelshaus. The panel found no significant health risks associated with the use of 2,4,5,-T and recommended a repeal of the limited ban.

The report was leaked to a group of scientists and consumer representatives before it was formally released and before Ruckelshaus could act on the recommendations. A predictable furor arose. The environmentalists issued a highly critical review of the report's methodologies and conclusions.[29] The EPA quickly reviewed the evidence, consulting with scientists from the Food and Drug Administration who had done the earlier tests on 2,4,5,-T with laboratory animals. The FDA scientists agreed with the critics of the leaked report and recommended that the ban be continued. Ruckelshaus thereupon in August 1971 rejected the advice of the ad hoc panel and maintained the partial ban on 2,4,5,-T.

The FDA scientists had been convened informally like the panel assembled under Mrak's chairmanship. Similarly, they conducted their deliberations without public hearings (the Federal Advisory Committee Act was not yet in existence). The decisionmaking process of the Department of Agriculture, and afterward of the EPA, first produced and then sustained a limited ban on the use of a herbicide thought to pose a human health risk when used as a weed killer in the vicinity of homes. On the face of it, this appeared to be a rational and responsible outcome. The authorities acted responsibly in the light of the facts of the case. But this is not how the episode was portrayed by critics of the EPA.

The Mrak report, which was condemned as biased by the environmental activists, was merely one review using one methodology to assess the scientific evidence. The EPA did not intend to rely on this report as the sole basis for decision and did not in fact do so. EPA officials decided that they should consider all the available evidence on the hazardous potential of 2,4,5,-T before making a decision. The EPA also decided that the report of the Mrak panel would be released to the public for debate and scrutiny. Again this appears to have been a reasonable position when viewed two decades later. There was no secrecy because the report, once completed, was to be made public in a timely fashion. But in the political heat of the moment critics were concerned that secrecy was distorting the whole policy process and that the main flaw with the Mrak panel was that the work was done without public involvement.

Mrak in fact offered to consider in his committee's review any epidemiological evidence the environmentalists wished to submit. He believed, however, that the environmentalists did not wish to submit research results but to "advocate their position."[30] Denied access to advocate their position (but granted access to submit evidence), the critics denounced the proceedings as being engineered to avoid public scrutiny and involvement. While the environmentalist critics sensed the public mood and gained a political victory, their attacks on the agency and the Mrak panel were clearly unfair. And curiously the lesson drawn from the episode was that the decisionmaking process was seriously flawed. As one critic wrote, "The history of the 2,4,5,-T episode is cogent evidence of the shambles into which the official decisionmaking machinery has lapsed. . . . The intervention of outside methods has been essential in keeping the government machinery on the rails. . . . In short, the established machinery for protecting the public health has failed, and failed ignominiously."[31]

Many of the key participants drew the conclusion that the way to prevent future disputes was to open the decisionmaking process to public scrutiny. As Ruckelshaus declared in a September 13, 1971, speech to the American Chemical Society:

I am convinced that if a decision regarding the use of particular chemicals is to have credibility with the public and with the media who may strongly influence that public judgment, then the decision must be made in the full glare of the public limelight.

It no longer suffices for me to call a group of scientists to my office and when we have finished to announce that based on their advice I have arrived at a certain decision. Rather, it is necessary for me to lay my scientific evidence and advice on the table where it may be examined and, indeed, cross-examined by other scientists and the public alike before I make a final decision.[32]

The corollary to the need for openness was the importance of independent scientific advice. Technical advice per se was not lacking within the EPA. As of October 1971 there were more than twenty technical advisory committees and various informal committees serving the EPA divisions (mostly holdover boards from the agencies transferred to the EPA).[33] A pesticide advisory group had been created as early as 1908 in the Department of Agriculture. The problem, according to the conventional wisdom, was that these groups were too closely tied to the industries and constituencies they served. Moreover, they had grown intellectually inbred and were out of touch with the latest scientific developments.

Not everyone shared the view that openness and independence were obvious answers to the generally acknowledged weaknesses of the advisory committees. Thomas E. Carroll, assistant administrator of the EPA, testified before Senator Metcalf's Subcommittee on Intergovernmental Relations of the Senate Committee on Government Operations in October 1971 that "opening all of the deliberations of these technical committees to the press would inhibit the development of the scientific background and data that go into the report."[34] Moreover, there was little danger, in Carroll's view, that the EPA would be captive to bad advice from any one source, since it routinely drew on multiple sources of advice, including the EPA's own technical staff. The EPA could and would ignore biased or inadequate technical advice.

Debate about public access to "predecisional" matters (the exception five of the Freedom of Information Act) also surfaced in the testimony. Carroll urged the subcommittee to judge outcomes, not process. In his view efforts to define how executive agencies should operate merely interfered with flexibility and did not achieve the desired ends.

The testimony by various executive branch witnesses before Senator Metcalf's Subcommittee on Intergovernmental Relations in October 1971 was a fascinating exploration of the central issues of "open government." In the numerous courteous but candid exchanges that took place, one senses the nation moving toward a turning point. The idea of the Federal Advisory Committee Act begins to take shape. Senator Metcalf and his colleagues who advocated openness in government believed that any lasting impact had to come from changing the way government operated. They were evidently not convinced that legislation was needed to enforce open meetings prior to the 2,4,5,-T case. That episode, however, appears to have crystallized and brought into focus Metcalf's views on the need for statutory regulation of federal advisory committees.[35]

Whether they were convinced outside scientific advice was necessary or believed public confidence could be restored only by a more transparent process of scientific advice, Ruckelshaus and his close colleagues embraced notions of the importance of openness in government similar to those of Metcalf. Neither the EPA leadership nor the Nixon administration raised serious objections to the passage of the FACA a year later. President Nixon signed the bill into law on October 6, 1972.

Ruckelshaus and his colleagues did not believe, however, that advisory committees should be specifically created by statute or that their terms of reference should be defined by statute. One problem with the early EPA advisory system was the clutter of committees, some of them mandated by law, that had come along with the functions transferred to the new agency. Many of these committees had a narrow charter or membership. Like other executive agencies, the EPA preferred the flexibility of committees that it could reorganize or change as new needs arose. The FACA, the EPA believed, was a generic statute mandating openness but not prescribing specific committees for the agencies. The FACA would, it was hoped, lessen detailed congressional intervention in return for the executive branch committing to a more open process.

The SAB Takes Shape

While the FACA was debated and eventually passed by Congress, the EPA began to move toward revamping its science advisory system to increase its independence from industrial sources. Four of the committees (of the more than twenty inherited at the EPA's creation) were abolished, three new ones were added, and several were refocused on broader mandates. EPA negotiations with industry over lead standards in gasoline were opened to public scrutiny, and independent scientists were invited to evaluate and comment on the accumulated evidence on the health effects of lead exposure.[36] All reports from advisory committees were to be made public simultaneously with transmission to the EPA.

The idea of a new, agencywide Science Advisory Board also began to take shape. Stanley M. Greenfield, the official brought in as the EPA assistant administrator for research and monitoring, came from the Rand Corporation in Santa Monica, California. Greenfield had served while at Rand on the Air Force's Science Advisory Board. The independent advice of this board, in his view, had been highly valuable to the Air Force. He initiated discussions with his colleagues about a similar group for EPA to provide outreach to the wider technical community and advise on the agency's scientific programs.

In July 1972 Greenfield circulated an information memorandum to the assistant administrators and deputy administrator calling for the establishment of a science advisory board for the EPA. "It seems most desirable and timely," the memo concluded, "to establish a cohesive structure of science advisory committees for this agency through the formation of an integrated Science Advisory Board."[37] The envisioned board would draw together "a few dozen prominent scientists from academia and industry" into a set of committees that would "reflect the major technical areas of EPA's concern."[38] The committee structure would consist of established advisory groups already in existence and serving the agency, special ad hoc panels set up for temporary periods, and possible new standing committees. The functions of the board would include advice on scientific matters and in-depth studies of wider problems drawing on the board's specialist skills.

On August 15, 1972, Greenfield followed up with an action memorandum calling for a decision on the board's creation. The memorandum added the argument that the SAB would provide "better routes to acquiring extra-mural advocacy for the EPA's technical stance by the national scientific community."[39]

Greenfield's proposal encountered substantial resistance from his fellow assistant administrators. The opposition ranged from minor bureaucratic and turf objections to profound conceptual concerns over the relationship of scientific knowledge to policy. As in the struggle to establish the Defense Science Board, the issue arose repeatedly of how to reconcile the aims of science with the agency's program goals.

Greenfield's colleagues asked, for example, why it was necessary to have a central board set up under the purview of the assistant administrator for research when in fact the operating divisions were in need of scientific advice geared to their own programs. John Quarles, an assistant administrator, argued that the "subordination of SAB to [the] chief scientist [is] not proper here because EPA [is] more tech-

nically involved throughout than other agencies."[40] If the existing
committees were absorbed into a new central board, the EPA operating units would be deprived of the immediate expertise they needed.
At a minimum the expertise would be diluted and less directly useful
to the operating units.

What seemed self-evident to Greenfield—that the SAB should be
located in the research bureau, since it was a committee of scientists—
was the point most seriously disputed by other assistant administrators. They were interested less in advancing the agency's research base
than in strengthening a major program goal, the capacity to issue timely
environmental regulations and standards. Technical uncertainty seemed
to them an inescapable concomitant of regulation. Operating officials
believed that science could not provide unambiguous answers to the
manifold uncertainties and that technical experts disagreed constantly
on methodologies, interpretations of data, and future needs. Therefore they could not see how adding more resources to the EPA's science and research base would make the agency more credible. Critics
would always find something to quarrel about or could always find
some scientist to challenge agency methodologies. The important point
to officials was that scientific advice be available to them at the time
they needed it and in a form they could use. The best science in their
terms was what was useful and not necessarily what was judged best
by the academic peer review process. Moreover, at the point of decision on regulations, science was only one among many factors that
came into play.[41]

Faced with the opposition of his colleagues, Greenfield was forced
to compromise. He felt strongly that the SAB concept had to be sal-
vaged and that the board in some form was critical to the agency. As
he later recalled:

> I complained to Bill [Ruckelshaus] at that time that this was going
> to haunt us later on, that we didn't have the data, that we were
> facing large uncertainties, that we had problems that would not go
> away in terms of acceptance by the scientific community . . . par-
> ticularly if you got to a point where the industry got smarter, and
> was able to use the uncertainties to fight some of the standards.[42]

Ruckelshaus backed Greenfield in going forward with the SAB, but
insisted on continued vigorous pursuit of enforcement goals while the
agency strove to bolster its scientific image. In the compromise plan

Greenfield agreed to separate the grants review process from the SAB, assured other assistant administrators that only those committees that wished to affiliate with the SAB need do so, and added an executive committee for the SAB made up of the chairmen of existing EPA committees plus a small number of other eminent outside scientists. Greenfield invited Mrak, now chairman of the Hazardous Materials Advisory Committee (the renewed and broadened committee that had assessed the scientific evidence on the herbicide 2,4,5,-T), to serve as chairman of the SAB's executive committee. Greenfield circulated the compromise plan together with the announcement of Mrak's appointment in a January 1973 action memorandum to his EPA colleagues. Finally, on June 26, 1973, the acting EPA administrator, Robert W. Fri, in Administrator's Decision Statement No. 6, officially created the SAB. Its chief functions were to

—provide scientific advice directly to the administrator;

—respond to specific issues raised by any agency component;

—evaluate the scientific needs of all components of the agency and periodically report on them; and

—evaluate the adequacy of its research programs in meeting identified needs.[43]

The statement identified the seven committees making up the SAB and outlined their roles. In September 1973 Thomas Bath, a Ph.D. chemical engineer from the University of Michigan who had worked in the EPA's Office of Research and Development since 1971, was named SAB staff director. He filed a charter for the new board with the OMB the same month and issued a new paper on the "function and mission" of the SAB. Health and energy advisory committees were added in the future. Bath secured the concurrence of the OMB for the SAB in January 1974 and informally reviewed the plans for the board with key staffers on Senator Muskie's Committee on Public Works.

On January 27, 1974, the executive committee of the SAB held its first formal meeting. The new EPA administrator, Russell Train, spoke to the group and pledged the EPA's close cooperation with the scientific community. His charge to the board typifies the decisionmaker's view of the collateral purposes served by the advisory body: "It is exceedingly important that we reach out and communicate effectively with all of the elements of the scientific community. . . . It is exceedingly important that I have at least one source of scientific advice and support that is not internal."[44]

Early History

The early history of the SAB was a replay and continuation of the difficulties and internal tensions that arose during its creation. Much of the early period was spent in squabbles over the board's mission, committee structure, organizational location, and style of operation. Periods of inactivity were interspersed with intense but often inconsequential quarrels between the Office of Research and Development (ORD) and other EPA offices.

Almost from the start, the board's location within the ORD was challenged as inappropriate. This location made the board focus too exclusively on research priorities and not enough on policy issues of interest to the operating divisions. In September 1974 Alvin Alm, assistant administrator for planning and management, launched an attack on "the inactivity of the SAB . . . and the absence of operating procedures," and called for the board to be transferred from ORD to the Office of the Administrator.[45] The ORD fought the proposed transfer, citing the need to protect important SAB efforts to review the EPA's overall research strategy for fiscal 1975. Administrator Train decided on March 6, 1975, to approve the transfer in the belief that this would enhance the board's effectiveness agencywide and affirm the broader definition of its mission.

Another nettlesome issue was the SAB's committee structure. The board initially had only the collection of committees that Greenfield could assemble by persuading his fellow assistant administrators. He recognized from the beginning that a restructuring of committees would eventually be necessary. Mrak, Bath, and assistant administrator Wilson K. Talley favored restructuring the committee system along disciplinary lines. The theory was that the existing structure was haphazard, mostly the result of evolution in the responsibilities of the old bureaus. A new disciplinary structure would provide a roster of highly competent scientists who could be assembled as needed into a matrix of subcommittees to work on problems cutting across jurisdictional lines. The plan was to create three new committees reflecting this approach and then to gradually terminate the old committees, transferring in the process some existing committee members to the new committees (the new committees were set up in June 1975).

The problem was that the disciplinary committee lines tended to harden into rigid boundaries of their own, making it more, not less, difficult to address practical problems. As one critic put it, "The problems don't come packaged by discipline. The problems are inter-disciplinary."[46]

Committees organized essentially along disciplinary lines did well with the traditional peer review of laboratories, for example, but less well with the new regulatory science issues the agency was facing. The problem, in brief, became the EPA variant of the issue faced by the Defense Department in the early phases of the Defense Science Board: how to get scientists to focus on the problems of concern to the agency rather than on the development of science itself. Regulatory science was gradually seen to be different from academic science.[47]

The SAB's initial style of operation in dealing with the agency also came in for criticism. In one of the board's early reports, it sharply criticized the EPA's handling of a research program known as the Community Health and Environmental Surveillance System (CHESS), a carryover project from the National Air Pollution Control Administration of the Department of Health, Education, and Welfare. The 1975 report attacked the epidemiological methods and data presentation contained in a CHESS monograph on sulphur oxides. A series of exposé articles on the CHESS study published in 1976 in the Los Angeles Times compounded the problem, citing the distortion of data to buttress agency regulatory policies.[48]

Two congressional committees thereupon held joint hearings to investigate the charges. The congressional investigators exonerated the EPA staff members from the allegations of fraud but found managerial shortcomings in the CHESS program and enough technical errors to render the monograph on sulphur oxides useless as a basis for regulatory action. The SAB, for its part, was criticized for providing to the EPA "only formal criticism, expressed in public meetings attended by press and industry."[49] The congressional report concluded that the relationship between the EPA and the board should be less formal, less adversarial, and more forward-looking rather than retrospective and critical.

Meanwhile, on a practical level, the SAB's role vis-à-vis that of the internal technical staff was a subject of confusion and uncertainty. Bath, the board's first executive director, decided at the change of administration to resign, enabling President Carter's newly appointed EPA administrator, Douglas Costle, to choose his own SAB staff director (though Bath had been appointed as a career official). Costle chose his long-time associate Richard Doud, whom he also named his science adviser. Doud shared Costle's objective of shifting EPA toward a focus on human health issues. His easy access to Costle as science adviser and their close working relationship raised the issue of whether the SAB should function as a staff to him or remain independent. The relationship of the SAB to the planning and analysis division headed by assistant administrator Steven Gage was problematic.

The need for clarifying the role of its scientific advisers was an important factor behind an EPA request to the National Academy of Science for a comprehensive review of the agency's decisionmaking procedures. The 1977 NAS report recognized the need for the EPA to bolster its scientific capacities for both technical and political reasons. The SAB could play an important role, the report concluded, in helping to detect errors, biases, and inadequacies in the agency's approach before policy conclusions and binding regulations were formulated. Changes were also suggested in the board's operations, including the merger of the EPA science adviser office with that of the SAB. The role envisaged by the academy for the board seemed to be that of a neutral umpire between the agency's tendency to act as advocate and industry's use of information to bolster its own position: "Much of the process by which EPA makes regulatory decisions is adversarial, and often scientific information is provided by one of the principals. Similarly, the agency itself is sometimes placed in an advocate role. In either case, review can help to assure a balanced treatment of scientific and technical information."[50]

The academy, though implicitly recognizing that regulatory science had political dimensions, did not fully come to grips with the potential conflicts between the adviser's roles as dispenser of purely technical information and as participant in the policy process. The NAS report recommended, for example, that the term of the SAB chairman be kept short, preferably not more than two years, so as to preserve the chairman's independence from the agency. This did not mesh with the thrust of the recommendation for the board's early involvement in EPA technical review to remedy error before irreversible commitments were made. If the board were an intimate adviser from the start, its own neutrality would presumably be compromised. To avoid this, however, one would not want the board to move back toward adversarial and overly formal procedures. But the balance to be struck between acting as a scientist and acting as a scientist-adviser intimately familiar with the agency's problems was not fully addressed in the NAS report. This central dilemma of regulatory science can never be definitively resolved. There is no permanent solution to what is inevitably a process of balancing legitimate but conflicting objectives.

Congress was unwilling to let the EPA work out its own science advisory structure. In 1977 it mandated a formal review of air-quality criteria documents by the SAB Clean Air Committee; in 1978 it broadened this review function to a wide range of agency technical activities, and gave statutory footing to the board.[51] Doud and other EPA officials felt uncomfortable with the specific mandates for SAB review of agency operations, but chose not to fight the congressional action because the statutory foundation ensured the board's continued existence. At that time the agency was engaged in an intermittent cold war with the OMB over the number of its advisory committees. The OMB was in its aggressive posture toward advisory committees and put intense pressure on Doud to abolish or consolidate committees. The 1977 and 1978 congressional actions removed the need for the SAB to battle further.

As the SAB was drawn increasingly into the central regulatory issues facing the agency, it came to be viewed as a potentially useful ally by the principal actors in the regulatory process. Virtually none of the major interests, whatever their beliefs about good science, were willing to accept the simple distinction between science and politics. Each strove to influence appointments to the board and interpret its findings in a favorable light. The pressures erupted into open conflict in the first Reagan term. As Jasanoff observes, "Attempts to control appointments to the SAB became one of the mini-arenas on which pro- and anti-regulatory interests played out their deep-seated philosophical and social conflicts."[52]

The First Reagan Term

The first Reagan term was a particularly frustrating time for EPA science advisers. Memories of the events are now overlaid with a great deal of folklore, emotion, and conventional wisdom—and interpretations are heavily influenced by political predispositions. The fundamental outlines, however, seem clear: the science advisory system was singled out for particular critical scrutiny; it was downgraded and went into near eclipse as it bore the brunt of Gorsuch-era political battles; but after 1983 it was resuscitated and set on its present course. The "hit list" episode, which came to light in 1983, marked both the low point in the SAB's fortunes and the beginning of its upturn toward the prominent status it currently enjoys.

Environmental politics was not among the top priorities of President Reagan and his advisers when they assumed office in January 1981. Fiscal policy and national security issues dominated their agenda. Consequently the environmental area was left largely to the interplay of, on the one hand, an ideological wing of Reagan Republicans in the policy-level positions and, on the other, the career officials who sought to temper the deregulatory impulse. (The Carter period had been marked by the reverse phenomenon: the career bureaucracy sought to moderate a militant regulatory thrust by the Carter appointees.) Although the White House–level Regulatory Review Group, chaired by Vice President Bush and staffed by the OMB, exercised a loose check on agency initiatives, the environmental team was left wide latitude to formulate policy.[53] So long as the administration was not embarrassed the EPA would be left alone. Unhappily this state of affairs did not last long.

The new appointees took the view that the SAB had become populated with environmental and science-for-the-people advocates during the Carter administration. The departing EPA administrator, Costle, apparently at least partly agreed; he had come to feel that the board was too large and needed a housecleaning to upgrade its scientific status and its usefulness to the agency. He had begun not renewing some board appointments. The Reagan officials now put a hold on all appointments to the SAB pending a policy review. For most of 1981 the SAB was inactive. The new administrator, Anne Gorsuch, and her team pondered the board's role but never came to a clear decision.

From the assumption that environmental radicals filled the ranks of EPA science advisers, some Reagan appointees inferred that scientists in the agency were generally a source of trouble. Accordingly the growth of the EPA's internal scientific work force began to be curbed. According to a study by the General Accounting Office, the EPA's scientific and engineering staff declined by 6.2 percent between 1973 and 1984, largely because of agencywide controls on hiring imposed by the Reagan administration.[54] The EPA's nonscientific work force increased by 9.4 percent during the same period, making the agency the only one of seven surveyed by the GAO in which the number of scientists declined while the total work force grew.

During this period industry advocates of "good science" were advancing a proposal to centralize the regulatory scientists from several agencies into one laboratory. It would presumably be better equipped and managed than the various laboratories then serving the regulatory agencies. This proposal had some virtues, but it could also be seen as working in parallel with the internal curbs on the scientific work force. As one industry representative said to me at the time, only half in jest, "First we will centralize them, and then abolish them."

As the political controversies intensified around the embattled Gorsuch (now Burford), science's role within the agency and the role of outside science advisers came in for intense debate. The science advocates were not without political skills and resources. They kept up a media campaign and fought to keep the advisory structure intact. As one former EPA official described the strategy, "There was an informal network of the old science advisers who kept in touch. We did a pretty good job of keeping her [Burford] off balance. Every time she tried to turn around we hit her with unfavorable publicity."[55]

Under a new executive director, Terry Yosie, the SAB kept up some activity even when the committees scarcely met and the appointment process was bogged down. Parts of the business community were important allies in focusing attention on the need for improving the agency's scientific capacities.

Environmental interest groups were also keenly interested in the SAB, though their perspective often diverged from that of the scientists. They veered at times toward a science-for-the-people approach, stressing the need to balance scientific with lay views and values. In 1982 environmental groups launched a successful initiative in Congress to diversify the SAB membership. A rider attached to an EPA appropriations bill for R&D funds stipulated that the board include representation from "states, industry, labor, academia, consumers, and the general public," justifying such interest balancing by reference to the FACA provision for committees to be "fairly balanced in terms of points of view represented."

The White House strongly objected to such populist influences in the environmental policy process. President Reagan vetoed the measure just before the 1982 congressional elections with a strongly worded message affirming the political neutrality of science: "The purpose of the Science Advisory Board is to apply the universally accepted premise of scientific peer review to the research conclusions that will form the basis for EPA regulations, a function that must remain above interest group politics."[56]

The president's stand received strong support from high echelons of the nation's scientific establishment when Frank Press, president of the National Academy of Sciences, and William Carey, executive officer for the American Association for the Advancement of Science, repudiated the effort to make the EPA's principal advisory committee the focus of explicit interest group representation.

The president's strong rhetorical stand on behalf of the political neutrality of science, however, turned into acute embarrassment in 1983 when the hit-list episode came to light. The House Committee on Science and Technology made public a document compiled by unknown sources in the early days of the Reagan presidency listing some ninety scientists from various EPA advisory boards. Reminiscent of the Nixon enemies' list, the document contained references of the sort that made for a press field day. Scientists were referred to as a "snail-darter type," a "Nader on toxics," or administrators were simply admonished to "get him (her) out."[57]

The disclosure constituted a serious embarrassment for the administration, evidently contradicting the president's commitment to scientific integrity in the regulatory process. Whether the hit list actually had much impact or was evidence of a concentrated effort to purge the SAB is uncertain. The entire episode has been befogged with misrepresentation, official denials, and exaggeration by critics. But scientists given negative ratings were generally not invited back to advise the agency. Whatever the motivation behind the list, the resulting publicity focused attention on the board and its role in the regulatory process. From a posture of decline and inactivity, the SAB now entered into a new phase of growth in size, level of activity, and overall importance to the EPA. Though often cast in terms of restoring scientific integrity, the subsequent recovery of the SAB can be more accurately understood as a case of the board's leadership quietly coming to an accommodation with the agency leadership.

Maturity: Gains in Status and Influence

Following the negative publicity of the hit list, the EPA began to devote new resources to the board. With the departure of Burford and the return of William Ruckelshaus as EPA administrator in March 1983, the board entered on a period of significant expansion in the number of its reports, program reviews, technical analyses, reviews of scientific methodologies, advisory documents, committee meetings,

Table 4-1. **Growth of the EPA Science Advisory Board, 1982–88**

Year	Members	Costs (in dollars)	Staff	Number of reports
1982	31	498,318	9	5
1983	37	581,243	8	8
1984	41	892,021	13	15
1985	54	1,050,270	12	38
1986	53	1,027,000	12	24
1987	67	1,186,966	12	29
1988	67	1,161,500	14	43

Source: Sheila Jasanoff, *The Fifth Branch: Science Advisors as Policymakers* (Harvard University Press, 1990), p. 90; and updated material from the SAB.

and informal consultations. Table 4-1 summarizes the highlights of the SAB's recovery and growth.

The EPA recovered full authority to make appointments to the SAB, and in essence depoliticized the appointment process. Vacancies were announced in the *Federal Register*; and a search for respected scientific experts, principally from the universities, without regard for their political affiliations or viewpoints, has since been the norm. Politics in the more subtle sense of balancing different scientific points of view, disciplines, and methodological orientations certainly remains. A premium has also been placed on familiarity with the agency's problems and procedures rather than purely academic standing in the choice of SAB members.

With Ruckelshaus's return came a more open style of decisionmaking and a shift in the agency's overall tone. The new team did much to restore agency morale and public confidence. Beyond this, an important policy thrust for Ruckelshaus was to deal with the issue of the EPA's scientific credibility. This was an issue transcending the hit list. It constituted one of the fundamental problems that he had sidestepped in initially establishing the strong enforcement orientation for the EPA.

Taking his cue from an influential 1983 report of the National Academy of Sciences, Ruckelshaus distinguished between the "risk assessment" and the "risk management" process.[58] Risk assessment was the whole set of procedures of research, testing, and experimentation that would define, characterize, and assign quantitative dimensions to science and environmental risks. These activities were to be conducted insofar as practicable in accord with strict scientific standards and would be separated from the agency's enforcement activities. Then risk management—the effort to enforce regulatory measures once risks had been accurately identified and characterized—would take over. Risk management activities would be kept institutionally and conceptually distinct within the agency (just as, for example, a so-called Chinese Wall might be maintained between different functions within a firm).

This separation could not and would not be total, because there was an inevitable overlap between the categories. As one moved further along the complex chain of criteria documents, thresholds, and standard setting for the various categories of risks, the processes of science and of regulation would become intermingled. The opinions of the agency's science advisers were of course required at critical points. Though the distinction thus could not be absolute, the concept of risk assessment and risk management as analytically distinct processes was an advance in the level of sophistication that the agency brought to its mission. It was a useful formula for institutionalizing the political neutrality of the SAB and other advisory committees. It gave the agency, at least for the critical post-hit-list period, an operational code that successfully wedded the science advisory effort to the regulatory process.

Possibly as a result of the renewed attention to the regulatory science issues, the EPA over the next several years resolved or partly resolved a number of knotty problems. In September 1986 the EPA issued the final version of Guidelines for Carcinogen Risk Assessment after nearly a decade of acrimonious debates among the agency, the chemical industry, the scientific community, and the OMB.[59] A SAB review of epidemiological and pharmacokinetic data assisted the EPA in the issuance of a final risk assessment on formaldehyde in April 1987.

The enormous internal efforts and multiple outside peer reviews involved in such success stories should, however, temper any easy optimism about the methodological gains. The parties engaged in the longstanding disputes may have simply reached a point of exhaustion and finally acquiesced to peer review in order to avert further court action. Several European nations evidently resolved similar disputes earlier while expending fewer resources.[60]

The sometimes almost obsessive focus on the scientific aspects of the agency's work has appalled some critics. They have contended that the EPA has been locked in a series of technocratic scientific disputes that amount to "asking the wrong questions."[61] In their view a scientific logic has displaced the more appropriate concern with preventive measures and with low-cost compliance, transmedia environmental phenomena, and other central policy issues. In part, the criticisms reflect the longstanding methodological division between ecologists and molecular scientists.

Ruckelshaus in any case later backed off from his risk assessment formulation and began emphasizing other aspects of the EPA mission. His new priorities included prevention, delegation of responsibilities to the states and localities, and a shift away from the preoccupation with human health issues. In these new emphases Ruckelshaus began to move the agency toward the changes that marked the terms of his successors—Lee Thomas, who became administrator in 1985, and William Reilly, who took office in 1989.

The Contemporary SAB

In 1986 Thomas initiated one of the most ambitious agency efforts to date: a comprehensive look at the entire range of the EPA's activities combined with an effort to establish agencywide priorities among them. Directed by Richard Morgenstern of the Office of Policy Analysis, and drawing on more than seventy-five of the agency's professional staff, the effort produced a much-discussed 1987 report, *Unfinished Business*. This report tried to determine for the first time which problems posed the greatest risk to human health and to the environment.[62]

The task proved to be a methodological nightmare, given the inadequate data, uncertain techniques, and trade-offs among uncommensurable objectives that it sought to devise. The team emerged with a ranking of thirty-one problem areas based on its best judgment and the available evidence, without pretending that the list represented scientific certitude. The team's assessment of the most serious environmental problems was strikingly at odds with public perceptions. Further, the allocation of agency resources reflected this mismatch between real and perceived problems. For example, many of the things the public was most concerned about and the EPA was devoting vast resources to—such as hazardous waste and underground storage tanks—posed relatively small risks, while other critical problems such as global climate change were being virtually ignored.[63] In 1987 the agency was spending several billion dollars for hazardous waste cleanup and only several million for climate change and indoor air pollution.

The obvious implication of the study was that the EPA should substantially reorder its priorities and budget allocations and should begin to educate the public. *Unfinished Business* had a substantial impact on thinking both within and outside the agency; but its effects on agency practice were, initially at least, much more limited. Not the least of the problems was the perception within the agency that the study was itself unfinished. In particular, critics saw it as a relatively crude, unscientific, first approximation of a complex reality that was not yet a sufficient basis for policy change.

In the spring of 1987 EPA administrator Thomas undertook a related initiative to shift agency priorities. He commissioned SAB's Research Strategies Committee to review the EPA's R&D program and to recommend appropriate changes in the types of research undertaken in order to serve more effectively the agency's long-term policy objectives.[64] The committee's report, *Future Risk*, issued a year later, became something of a best seller. More than 5,000 copies were printed and distributed within the first year of publication.[65] The committee's recommendations, though ostensibly directed toward research strategies, ranged well beyond a narrow concern with R&D to fundamental strategic and policy concerns. All of the committee's specific proposals relating to R&D were cast, for example, within a conceptual framework favoring "a strategic shift in emphasis from control and clean-up to anticipation and prevention."[66]

When William Reilly took office as President Bush's EPA administrator, he began with a strong commitment to the approach reflected in the *Unfinished Business* and *Future Risk* reports. He told close associates that he had two main priorities: to convince the public and Congress that the administration was seriously committed to environmental goals, and to pick up and carry forward the reordering of priorities begun with *Unfinished Business*. Accordingly he charged the SAB as one of his first official actions to review the findings and methodologies of *Unfinished Business*, applying the best scientific knowledge available to produce strategies for solving specific urgent problems or more broadly for mitigating a range of interrelated environmental risks.

The task proved to be a large challenge for the board, raising the issue of science versus policy and bringing divergent scientific perspectives into sharp conflict. The process of internal debate and intense activity set off by Reilly's request finally culminated in the September 25, 1990, release of a major new SAB report, *Reducing*

Risk: Setting Priorities and Strategies for Environmental Protection.
Reilly used the occasion of the publication of the report for an important address at the National Press Club outlining his views, fittingly
entitled, "Aiming Before We Shoot: The Quiet Revolution in Environmental Policy."[67] The speech and the SAB report largely endorsed
the findings and priorities of the *Unfinished Business* study, calling
for new attention to serious ecological hazards such as climate change
from greenhouse gases, to prevention rather than emission controls,
and to the need for improved regulatory science.

The SAB report was a major influence in legitimating Reilly's conception of the EPA's mission and helping him to change the agency's
organizational culture. The scientific aura and authority of the board
validated the policy shifts already in progress and gave Reilly and his
associates important support that they needed in dealing with internal
skeptics. The SAB's contribution was not in providing original advice
but in giving its stamp of approval to the policy directions in which
Thomas and then Reilly sought to move.

Reducing Risk proved to be the most popular report ever issued by
the EPA; more the 16,000 copies were distributed within six months
of its publication. For the board itself *Reducing Risk* represented an
important step, moving it inevitably into the center of EPA policy
debates. Although some SAB members continued to feel that they
were engaged principally in scientific rather than policy issues, the
effort clearly shifted the board toward a more general advisory role
and away from a strict focus on risk assessments of specific pollutants.

A conflict between the Health Effects Committee and the Ecology
Effects Committee was resolved with broad acceptance of the view
that both the EPA and the board should devote greater attention to
ecology and broad issues of habitat, climate change, and quality of
life. The assumption was that the agency could not neglect the large
environmental effects that could have profound long-term implications for human health, even though these effects could not be analyzed with the precision typical of biochemistry, toxicology, and other
laboratory sciences. Meanwhile the search for improved methodologies and for improved understanding of dose-response ratios and
quantitative risk characterization would continue. The more radical
critics of the EPA's scientific methods, such as Bruce Ames of the
University of California at Berkeley, who has assailed the agency's
focus on a narrow range of industrial pollutants and its neglect of
natural carcinogens, would also continue to spark a vigorous debate.[68]

The persons who played important roles in the SAB's *Reducing Risk* study and in subsequent intra-agency educational efforts illustrate the shift toward a more explicit concern with policy. Cochairmen of the study were Professor Raymond Loehr of the University of Texas (also the chairman of the SAB) and Jonathan Lash, until the fall of 1990 the secretary of natural resources for the state of Vermont. Others who brought considerable practical experience and an action orientation included William Cooper, a zoologist from Michigan State University and a long-time chairman of the Michigan Environmental Board, and Fred Hanson of Oregon's Department of Environmental Quality. Extensive briefings on the study were given by these scientists and other SAB members to large numbers of EPA staff at headquarters and in regional offices across the country. The ecological emphasis of *Reducing Risk* drew enthusiastic responses and support from the EPA staff; the study evidently both reflected and served to encourage the new concern with ecology within the agency.

In 1990 the EPA also achieved an important milestone in bringing to a conclusion a ten-year effort to reauthorize the Clean Air Act.[69] The culmination of a decade of controversy and executive-legislative struggle, this comprehensive environmental statute pleased both environmental groups and industry. It introduced novel economic incentives into the regulatory framework for acid rain rather than relying on a purely command approach. The Clean Air Act—even more than the Reagan era reauthorizations of the Toxic Substance Control Act and the Federal Insecticide, Fungicide, and Rodenticide Act and the Montreal Conventions—marked an important turning point for environmental politics. The 1990s may be a period when the policy gridlock can be eased and a more rational climate for debate restored. The EPA's scientific advisers have played a useful role in this whole process, legitimating and encouraging change initiated by the agency's leadership.

Epilogue

In the course of a decade the Science Advisory Board moved from a position on the periphery of EPA affairs into the center of the policy process. As Jasanoff has written, 'By the end of the 1980s . . . external indicators all suggested that the relationship between EPA and the SAB had settled into a pattern of successful cooperation. During the second Reagan term, in particular, the boards's resources and reputation grew, along with its breadth of involvement in EPA decision-making."[70] A few simple lessons may be drawn from this evolution of the board.

First, one is struck by the enormous hard work and persistent effort required before the board truly became useful to the agency. It seems almost banal to say, but success in the relationship between an advisory committee and the agency it serves does not come easily. It took nearly a decade and the accident of the hit-list episode to finally produce the resources for growth and the conditions for a receptive hearing from the agency.

In the field of regulatory affairs, the science adviser walks a fine line between assuming a technocratic, value-neutral stance (which brings with it the danger of being aloof and above the fray) and being a political partisan (which might mean being discredited as simply another partisan voice). To benefit from the aura of scientific detachment, while becoming well versed in the intricacies of agency policy, requires administrative and political skills of a high order. As with the Defense Science Board, the SAB cannot function in the utopian rationalist mode, since there are never purely technical answers to issues of regulatory policy. Yet perhaps even more than with the DSB, the environmental science adviser must partake of the objective aura of science even while being aware of the political realities behind the advice.

Second, the so-called independence that was so strongly a part of the original mandate for the board was a blind alley. The board was less successful when it cast itself in the role of an independent critic; it became successful only when it earned the confidence of the agency and moved into closer relationship with agency officials at all levels.

Third, the role of leadership within the advisory board stands out. It was only after the rotating chairmanship was dropped in favor of continuity that the board began to relate effectively to the agency. The chairman's internal role was notable in holding together a fractious body composed of forceful personalities and divergent professional and disciplinary backgrounds.[71] The chairman, by structuring the issues addressed and the recommendations, becomes the practical means to achieve the balance between political neutrality and policy relevance.

Fourth, without resources and support from the agency's leadership the advisory board is likely to wither on the vine. The interactions between adviser and agency must be broad as well as deep if there is to be a genuine impact on agency policy and operations. The

SAB became successful because first Thomas and then Reilly made it a crucial instrument of their policy approach and leadership style. Working relations were developed at all levels.

The SAB has become a successful example of the standing advisory committee that combines continuity and depth of expertise with flexibility and adaptiveness. Its main contribution has been not to provide novel and independent advice, but to provide validation and support to the agency's leaders as they have sought to foster a new culture and bureaucratic outlook.

5 ||| Advice, Policy Ambiguity, and Goal Confusion: The Department of Energy

THE TWO PREVIOUS CHAPTERS have shown how advisers have become important players in the policy process. In the Defense Department technical advice blends with complex political and institutional interests to shape national security policy. In the regulatory arena science advice has become an inescapable element in all aspects of policy and operations, even if the forces that ultimately shape the outcome are not the technical dimensions. I now move to cases where the uses of science advisers are much more ambiguous and problematical. The reasons for this will become clear as the case studies unfold. The advice reflects the lack of clarity and consensus in the agencies and in the wider political arena about the goals to be pursued.

Science advice can be productive when the agency is clear about its purpose and has a coherent sense of mission. Advisers can help to define the appropriate means to agreed ends and can clarify the ends when they are obscure or in dispute. But when there is no basic consensus on the agency's role, when the policy process is fragmented, or when the larger political system attaches low priority to the agency mission, the advisers function in a no-win situation. Advisers cannot produce a policy consensus in the face of deep political divisions and can only whistle in the wind if support for an agency is missing at the top levels of government.

The Department of Energy (DOE) and the National Aeronautics and Space Administration (see chapter 6) illustrate different aspects of the problems just mentioned. The DOE's fundamental problem has been lack of broad agreement over the government's role in meeting civilian energy needs. Some key actors want energy policy to be limited to tax policy, but they disagree over whether taxation should be used to promote conservation, raise revenues, or provide infrastructure. Others want energy policy to promote alternative fuels development, energy conservation, and new life styles, or to provide subsidies for particular sectors of industry.

Born out of the 1970s oil shocks and buffeted by the winds of deregulation in the 1980s, the DOE has never found a period of calm that would ensure its continued existence. Component bureaus have enjoyed strong congressional support, but coordinating policy among the department's fiefdoms has been all but impossible. And even if the executive branch could somehow produce a coherent blueprint for action, the dispersion of congressional power would guarantee that the parts remained more powerful than the whole.

Since the advisers have been forced to operate in this context, it is hardly surprising that they have had a hard time defining the central issues. Like the Defense Science Board and the EPA Science Advisory Board, the DOE's Energy Research Advisory Board (ERAB) started out by providing advice on relatively narrow issues of research policy (policy for science). But unlike its counterparts, ERAB was never able fully to break out of that mode. Science advice in the DOE was focused at the bureau or subagency level and lacked an integrative departmentwide perspective. The factions and the frictions within ERAB reflected these large disputes and made effective working relationships difficult.

Background

The Energy Research Advisory Board (ERAB)—and its successor, the Secretary's Energy Advisory Board (SEAB)—can be traced back to the period of the Atomic Energy Commission (AEC) from 1946 to 1974. The AEC's General Advisory Committee (GAC) during this time occupied a prominent and influential position. The GAC was created as part of the Atomic Energy Act of 1946, which established the AEC.[1] The GAC advised the commission on a wide range of issues relating to both civil and military aspects of nuclear energy. The H-bomb issue, the Oppenheimer security clearance controversy, and the decision to build a second weapons laboratory come readily to mind as major issues in which the GAC played a significant role. The GAC was the most important and most visible entity in a network of about twenty advisory committees serving the AEC in the early postwar period.[2]

With the abolition of the AEC and the creation of the Energy Research and Development Administration (ERDA) in 1974, the AEC's regulatory functions were split off into the Nuclear Regulatory Commission (NRC). The advisory committee took on a somewhat narrower mandate thereafter but continued to exist as a statutory advisory committee.

Congress's influential Joint Committee on Atomic Energy was also abolished in this reorganization, and its demise had much to do with the breakup of the powerful coalition fostering the development of nuclear energy (and enhancing the GAC's ability to exercise substantial influence). The split jurisdiction among ERDA, the NRC, and the White House staff, along with the demise of the powerful joint committee, caused a decline in the importance of the General Advisory Committee.

The disarray in energy policymaking following the first oil shock of 1973–74 continued into the next several years. It became evident that the creation of ERDA did not solve the problems identified by the critics of the AEC structure. A further reorganization became necessary. After protracted debate the split in authority between a White House-level energy adviser with quasi-operational responsibilities and ERDA as a weak independent agency came to be generally viewed as unworkable. The absence of cabinet rank for ERDA was seen as a related problem. The solution, so it appeared, was to create a cabinet-level department of energy.

On October 1, 1977, the Department of Energy was established. James R. Schlesinger, President Carter's energy adviser who had steered the reorganization through Congress, became the first secretary of energy. The GAC, which had been retained in the ERDA reorganization, was dropped from the legislation creating the new department. It is unclear whether this occurred through inadvertence or explicit congressional decision. In either case there was no serious effort to retain the advisory committee from either the scientific community or the circle of presidential energy advisers.

Early History

The Energy Research Advisory Board had its origins as part of the steps insisted upon by John Deutch, MIT chemist, prior to accepting the post of director of energy research and development in the new department. Deutch conceived of the board not as a re-creation of the old GAC with a wide-ranging mandate but as a more narrowly focused body dealing with energy research and development issues. He believed, however, that ERAB should report directly to Secretary Schlesinger. He felt it should respond to the secretary's needs for technical advice even if primarily focused on the energy R&D issues. Deutch went to considerable pains to find a chairman whom Schlesinger would respect and with whom he could work easily. Deutch and his staff chose Solomon J. Buchsbaum of Bell Laboratories, an experienced and widely respected scientist-administrator with broad experience in science advisory roles. He and Schlesinger were on cordial and respectful, if not personally close, terms.

Schlesinger was originally somewhat cool about ERAB (by nature he was not an admirer of consultants, advisers, or committees). A letter from his personal assistant to Deutch in 1977 bore a scrawled note at the top, "John—Do we really need this?"[3] Schlesinger initially referred to the board as "Deutch's committee," while Deutch characterized ERAB as "Schlesinger's committee." For approximately the first six months of ERAB's existence it had little or no role. It did meet several times but merely received briefings on DOE operations.

ERAB became significant when Schlesinger was faced with a specific problem on which he felt it could be of assistance—the mounting dispute over whether the University of California should continue to manage the Los Alamos and the Lawrence Livermore weapons laboratories. Faculty opposition was growing to the continuance of the university's relationship to the weapons laboratories. The Berkeley campus was especially militant. The critics' aim was to force the university to sever its ties with the labs.

When Schlesinger turned to ERAB for help, a subcommittee was convened consisting of Buchsbaum as chairman and ERAB members Hans Mark and Gerald Tape. The panel held a series of public hearings in 1978 and 1979 on the universities' laboratory ties. A report was subsequently issued saying, in effect, that the university was an appropriate body to supervise the laboratories and could continue to function effectively in that role. However, other arrangements could readily be made, should the regents of the University of California decide to sever ties with the labs.[4] The regents decided, outvoting California Governor Jerry Brown and his supporters on the board, that the University of California system should renew its ties with the DOE laboratories.[5]

For the duration of Schlesinger's tenure as secretary, with Deutch as director of energy research and development (and later under secretary) and with Buchsbaum as ERAB chairman, ERAB functioned reasonably well. With the departure of both Schlesinger and Deutch in August 1979, uncertainty set in, and the fortunes of ERAB took a turn for the worse. Buchsbaum stayed on as chairman until the end of the Carter administration, but the board did not regain the modest influence it had achieved.

Energy policy in the late 1970s was in a state of flux and disorganization. The second oil shock, of 1978–79, continued to focus attention on alternative fuels and conservation. But the support for expensive demonstration projects began to wane, and doubts persisted about the government's role in stimulating technology development. Demonstration projects had been a focus of energy policy since the first oil shock in the Nixon administration; this policy emphasis continued through the Ford administration and into the Carter term. The lack of effective commercial linkage led President Carter's science adviser Frank Press and the Office of Management and Budget, in a reorientation of budget priorities, to downgrade the demonstrations.[6] Basic research and civilian applied research began to receive more attention.

The political controversies surrounding energy policy that finally led to Schlesinger's resignation made it difficult for ERAB to find a coherent frame of reference for its activities. It struggled to develop a stable working relationship with Schlesinger and with the successor DOE leadership amid the conflicts, but it never fully succeeded.

Close working relationships take time to evolve. With advisory committees the period of adaptation to a leadership transition appears to be at least as protracted as with the internal staff (and probably more so since a new department head looks first to defining his or her relations with the regular full-time staff). The advisory relationship is less pressing and will normally await the clarification of other bureaucratic roles. The advisory committee can be handled by simply ignoring it for the moment while the new official attends to more pressing matters. ERAB was largely ignored after Schlesinger's departure.

One other development is noteworthy during this period. The Carter administration, despite its distrust of committees, did attach priority to making advisory committees more representative. ERAB was pressed to become more consumer and environmentally oriented. Environmental activist Amory Lovens was put on the board at White House insistence, to bring an environmental perspective. In principle trying to integrate environment and energy concerns was a highly desirable goal. In practice this particular blend of personalities proved to be disruptive. The lesson is that achieving balance in advisory committee membership should itself reflect a balancing of interests. A membership homogeneous enough to work together effectively must be balanced against a too like-minded body of individuals who merely reinforce one another's outlook. The personality factors, however, were probably in the end less decisive than the deep political conflicts that paralyzed energy policy.

The 1980s

The 1980s were a period of fluidity and dramatic shifts in energy priorities and in the fortune of the DOE. President Reagan had campaigned on a platform of abolishing the department because it interfered with the marketplace. Some controversies gradually waned, however, after the break in oil prices beginning in 1981. Deregulation of oil and natural gas was an early Reagan goal and was largely accomplished by the middle of Reagan's first term. By the end of the decade new controversies had arisen.

The need for creative thinking on a range of complex technical-policy issues was never in doubt, but the frame of reference for ERAB studies was always a matter of dispute within the department. ERAB's status, never an easy one since its inception, became increasingly problematical as the board struggled to keep relevant and abreast of policy shifts. Yet ERAB maintained a substantial pace of activity throughout the decade, turning out reports in a wide variety of areas. As figure 5-1 shows, ERAB activities centered on cross-cutting evaluations of R&D priorities, institutional relationships, nuclear issues, the research base underlying energy policy, and issues relating to DOE's national laboratories. A spurt of activity occurred at the start of the decade as many ongoing projects were brought to completion before the election.[7]

The inevitable change of priorities and personalities after the presidential transition left its mark on the board. The 1980 Republican platform was highly critical of the Carter administration's energy policies. Ronald Reagan's pledge to abolish the department in order to promote market forces naturally had a disquieting effect on departmental morale. Some Reagan priorities were virtually the opposite of Carter's: support for nuclear energy and aversion to solar research, whereas the Carter approach was hostile to nuclear energy and supportive of solar research. As the board sought to define its role, hold-over membership contributed to the disinterest toward the board on the part of the new DOE leadership. A close participant told me that at the start of the Reagan administration, ERAB was "paralyzed for nearly a whole year."

Gradually ERAB resumed an active existence, working through the team of Deputy Secretary Kenneth Davis and ERAB chairman Lewis H. Roddis, Jr. Its first report came on advanced isotope separation in May 1982; in September 1982 it produced a comprehensive review of the Department of Energy's multiprogram laboratories. As a very crude indicator of the level of activity, the board produced sixteen reports in the period 1979–81 and twenty-four reports in 1982–88.[8]

ERAB, however, continued to have difficulty in finding a comfortable fit with the top departmental leadership, in part because the leadership changed often. Moreover, it had few strong ties with the department's operating divisions. Severe DOE program cuts throughout the 1980s created an atmosphere in which ERAB could not easily operate. Everyone in the department was worried about budget cuts and afraid that any freewheeling study could be used to one's disadvantage in the perennial budget battles. As a result ERAB reports tended to be focused narrowly on the specific charge formulated in tasking letters. ERAB's secretariat seemed to be more concerned with preventing the board from being seen as a renegade than with mobilizing the board's creative energies. There were whole areas (notably the defense programs and the nuclear production facilities) from which the board was explicitly excluded. Most important, ERAB was not encouraged or permitted to initiate studies on its own. This is an acid test of whether a board has any short-run influence: if the board cannot initiate a study, it likely has no standing or stature (but if the advisers are too independent, they will not last long). To complicate the board's problem, there were three leadership successions in the department during the Reagan period. These led to inevitable readjustments and delays for ERAB.[9]

In the 1980s ERAB was, in short, functioning in an environment destined to make the adviser's life difficult. The ERAB mandate was either too broad or too narrow: too broad in that energy policy was a diffuse concept lacking clear boundaries and too narrow in that any particular study had its parameters very strictly delimited. The board had too little opportunity to exercise its creative judgment, and at

Table 5-1. **Overview of Studies by the Energy Research Advisory Board, 1980–88**

Area	1980	1981	1982	1983	1984	1985	1986	1987	1988
Cross-cut evaluation	R&D priorities			Federal energy role	Long-term guidelines			Energy competitiveness	
Institutional relationships		DOE labs	DOE-universities		DOE international			Education	Technical utilization
Nuclear issues	Advanced isotope separation	Advanced isotope separation	Light-water reactor safety		Light-water reactor R&D	Civilian nuclear power			New production reactor
Solar energy issues		Biomass	Solar...			Report solar power			
Conservation			Conservation						
Fossil					Clean coal		Solid earth sciences		
Research base			Materials R&D		Materials facility		Chemistry/physics		
Fusion research issues		P.L. 96-386.............		Fusion.................			Fusion		
Laboratory support	Los Alamos National Lab		- - - -Oak Ridge National Lab - - - -		- - - - - - - - - -Argonne National Lab- - - - - - - - - -				Pacific Northwest Lab

Source: Data from ERAB Secretariat, 1989.

other times it was set adrift in a vast sea with no clear guidelines. Budgetary constraints led to intradepartmental tensions and often caused the ERAB secretariat to withhold information from members of the board. Relations between the board and its staff could not be described as close or cordial.

An important variable is the degree to which the agency itself has clear authority to take program initiatives. An advisory body can be influential when its sponsoring agency is strong, has presidential backing (or at least acquiescence), has reasonably clear lines of authority and a supportive constituency in Congress, and has an interest group clientele that provides support. In that case, the agency can act on its own in a meaningful way, while nominally responding to presidential and congressional direction. When the agency itself is weak or cannot act in a given policy area because the White House or Congress has dominated policy and curbed the agency's discretionary authority, the advisers have little to say. And what they do say is irrelevant because the agency cannot act on the advice. Decisionmaking largely centered in the White House on energy issues in the 1980s, so the departmental leadership tended to be frustrated. What is perhaps surprising, given all these obstacles, is that ERAB was able to function at all in the 1980s.

ERAB made an indirect contribution in 1986 when at the suggestion of its chairman the secretary of energy turned to the National Academy of Sciences for a study of environmental and other problems affecting the nuclear production reactors. The NAS report found glaring weaknesses with the technical management of the department's nuclear programs for the military.[10] Recognition of the problems at the Rocky Flats and Savannah River facilities came slowly, however. The issue finally blossomed into a political crisis in 1988, and only then did it begin to receive serious attention. An advisory board with more leeway to initiate studies on its own might have been able to alert policymakers sooner to the impending dangers. But the problem in this case was less that of capturing the attention of executive branch policymakers than of capturing the attention of Congress and the mass media. The problem was to balance priorities between production volume and environmental goals, and to find the resources for the substantial new investment needed to modernize the nuclear weapons production system. As the cold war waned, it became evident that production of fissionable materials was less urgent and could be more appropriately balanced with environmental concerns.

Other DOE Technical Advisory Committees

While ERAB dealt with a range of difficult energy R&D issues, another tier of technical advisory committees addressed narrower and perhaps easier issues. One of the most successful was the High-Energy Physics Advisory Panel (HEPAP). Its mandate was to plan for and advise the secretary on issues relating to high-energy physics. HEPAP, since the early days of the AEC, has had no responsibilities beyond the health of high-energy physics as a scientific discipline. Over the years it has rendered important advisory service on the construction of new research facilities, the definition of research priorities, policies regarding the user group concept of operation, and related issues.

Four other DOE technical advisory committees somewhat akin to HEPAP were created in the 1980s. They have enjoyed varying degrees of success. These committees were the Basic Energy Sciences Advisory Committee, Health and Environmental Protection Advisory Committee, Innovative Control Technology Advisory Panel, and Magnetic Fusion Advisory Committee. They were created at the initiative of the director of energy research and development with the HEPAP model in mind.

These committees have been in general less successful than HEPAP for various reasons. Several have been interdisciplinary with a heterogeneous membership and a less clearly defined focus; their creation was inspired by public relations motives and the desire for protective bureaucratic coloration, and they have not been expected to do much. Nor have they had a large, well-funded program area such as high-energy physics that would attract and hold the interest of prestigious science advisers as committee members. None of the four smaller committees had any relationship to ERAB, unlike the umbrella-committee practices in the DOD and NASA. The Magnetic Fusion Advisory Committee was a temporary committee created for a special purpose: to map out a strategy for the future magnetic fusion effort.

The DOE also has had an assortment of other advisory committees, including (1) an Advisory Committee on Nuclear Facility Safety, which was mandated by congressional action, (2) a smaller committee with a narrow technical mandate, the American Statistical Association Committee on Energy Statistics, and (3) a classic illustration of the representative advisory committee, the National Coal Council, which speaks on behalf of a particular industry.[11]

Conflicts of Interest

In contrast to most other departments, the DOE has typically not classed its advisory committee members as *special government employees*. Thus its advisers have not been exposed to the conflict-of-interest laws and the potential criminal penalties in 18 U.S.C., the key statute. All DOE committee members are *representative* advisers, even though they appear plainly to fit the criteria that would require them to be classed as special government employees (that is, they are named by the agency, respond to a charge issued by the department or agency which created them, meet at the behest and under the supervision of the agency, and have a chairman designated by the agency).

Departmental legal officers have formally taken the position that none of their science advisers need to be treated as special government employees. The committee management officials have had no disposition to challenge the ruling. There is nothing in this practice that contravenes the Federal Advisory Committee Act, which places the major responsibility for oversight of committees with the agencies themselves. But the contrast with respect to conflict-of-interest practices suggests the need for clearer guidance, if only to affirm that the agencies have options as well as responsibilities in supervising their advisory systems.

ERAB and now SEAB advisers, however, are subject to financial disclosures. The practice follows the short-form recommendations of the Administrative Conference of the United States (see chapter 9). Members file a confidential statement with the SEAB secretariat.

In the DOE the onus to be alert to conflicts rests mainly with the committee members themselves. Since there are no criminal penalties of the kind that would apply when advisers are special government employees, the DOE advisers run no legal risks under the criminal penalties of 18 U.S.C. This looser interpretation of conflict requirements for advisory committee members contrasts with the department's very strict requirements on full-time DOE employees, which are much more stringent than in other executive departments. Yet this looser interpretation is more in keeping with the spirit of the conflict laws as they were originally drawn. The conflict laws did not originally intend that advisers were to be considered government employees or treated as such for ethics purposes. The stringent regulation of advisers seems to have been adopted by federal agencies in the wake of the 1978 Ethics in Government Act.

Energy Department counsel has ruled, however, that SEAB must comply fully with the FACA provisions on notice of meetings, the holding of open meetings, record keeping, and the rest. Indeed under the legislation setting up the DOE, the requirements for openness are more demanding than those contained under the FACA. Only national security considerations, for example, are a valid grounds for closing any DOE advisory committee meeting. This was the result of congressional suspicions of big oil and of the Natural Petroleum Council at the time of the DOE's creation.

Impact of Studies by the Energy Research Advisory Board

Few agencies have ever attempted to assess the impact of recommendations made by their advisory committees. But ERAB contracted with the Argonne National Laboratory for a study of the implementation of ERAB recommendations.[12] Eight studies were examined in a novel effort to assess what impact, if any, the studies had on department operations. DOE program offices were asked to comment on the usefulness of the ERAB recommendations and the extent to which they influenced departmental policy.

The report and the departmental comments were written in cautious bureaucratic language but seemed to echo a common theme, which one might term lukewarm praise. Several ERAB reports were characterized as being "generally useful" or "useful," and a few as "somewhat useful." Another was "helpful in reassessing priorities" but "only somewhat useful because of DOE budget limitations." In another case, the "recommendations . . . called for actions beyond the scope of influence of the Secretary of Energy." And, damning with faint praise, one study was said to have "received significant attention from DOE management."

Specific comments on detailed ERAB recommendations suggested that the department found many of them to be merely restatements of existing policy, others to have had little impact on policy and operations, some to be usefully supportive of ongoing activities, and others to be flatly wrong in the view of departmental officials. Only a few recommendations were found to be highly useful.

Comments on the overall quality of ERAB studies from the major units of the department were also lukewarm: "ERAB recommendations are just that, and not requirements"; "The Board needs to understand the practical limitations imposed on the DOE decisionmaking cycle created by the time lags of the congressional budgetary/planning cycle"; "The ERAB should limit its study to the specific charter requested in the Charge Letter . . . need improvement in Charge Letter so that a panel will have a clearer understanding of what it is being asked to do"; "ERAB members and panelists are not well-informed on DOE . . . additional iteration between ERAB panels and DOE program managers as studies are completed and recommendations being formulated may be useful."

Many small suggestions for improvement in the timing, distribution, and format of ERAB reports were made. But evidently the Argonne study itself was only mildly useful. None of its suggestions seemed to have been taken seriously. The major overhaul of ERAB undertaken by Admiral James D. Watkins when he became secretary of energy ignored the appraisal.

The Watkins Era

The problems that the department faced at the time Watkins became secretary in 1989 were formidable: an environmental disaster at its nuclear production facilities; once again the nation's growing dependence on imported oil; no clear national strategy on energy use or energy conservation; overly detailed congressional guidelines making comprehensive policy difficult; an extensive network of laboratories whose missions were in need of redefinition; unsolved problems of civilian nuclear waste disposal; and others. After a period of suspended activity during which Watkins conducted a review of ERAB, the science advisers were rechristened the Secretary's Energy Advisory Board on January 1, 1990. SEAB had a new charter and wholly new membership. Significantly the new board included defense issues as a major concern. ERAB had been confined to the narrower mandate of civilian energy technologies and the basic energy sciences as its main areas of activity. Some of the most significant department concerns in brief had been ruled off limits for ERAB.

When Secretary Watkins was contemplating the new role he envisaged for SEAB, he clearly intended that it fit with his style of operation and serve his priorities. The advisory system would inevitably reflect his unique blend of populism and technocracy. An early need in his mind was to get a firm grip on his own bureaucracy. The DOE was in his view especially obdurate, a collection of fiefdoms responding to narrow constituencies and a staff dominated by a technocratic logic. The populist Watkins determined that his department needed a shaking up.

Several early blasts by the secretary at his own department were unusual even for a town accustomed to seeing congressmen run against Congress and presidents attack the federal bureaucracy. Watkins's approach included a mixture of centralized decisionmaking and a radical thrust toward citizen participation. He determined that the department had been run by its separate domains—the military programs (with their powerful defense labs), fossil fuels, civilian nuclear energy, high-energy physics—for too long with too little central direction. But Watkins also wanted to open up the department's processes to the wider public. While he relied heavily on a circle of close aides, he made plans for an elaborate series of public hearings to gather reactions from the public on a national energy strategy. The public would speak and present ideas unconstrained by bureaucratic habits of thought.

Watkins's conception of the proposed science advisory board reflected the same attitudes. The reinvigorated board that emerged from Watkins's review had increased responsibilities, a different style of operation, and showed promise of influencing policy more than had its predecessor. Secretary Watkins made it plain that he would make full use of SEAB's advice and in particular that he himself would give the board its assignments.

He strongly favored outside advice, especially to offset the biases of the various departmental fiefdoms. The board would report directly to him in order to avoid getting caught up in lower-level jurisdictional fights, something he believed had happened to ERAB. All assignments and tasks would come from the secretary, with no intermediate layer or office to get in the way other than a small secretariat located in the secretary's immediate office.

Moreover, the new SEAB was to be a more representative body; and in particular, it should be composed of the social as well as the natural sciences, consumer and environmental representatives, and other public interest members. In the membership he included individuals representing environmental and consumer interests as a means of keeping the pressure on his own department to be alert to environmental concerns (about one-quarter of the members were nontechnical people, mostly environmentalists or representatives from public interest associations). Watkins wanted the board to operate informally when needed, without even having a fixed meeting schedule, to avoid becoming a process in search of a problem. And indeed he did not feel in a hurry to establish the new board. A more important priority was to begin the process of eliciting public views on the national energy strategy and reassuring the public. Popular advice was in a sense more important that expert advice. Environmental protection, he told close associates, had to be a primary initial concern of his stewardship of the department.

In his Senate confirmation hearings, Watkins asked Congress if he could defer for one year submission of the legislatively mandated national energy plan. The delay from April 1989 to April 1990 would enable him to submit a fully elaborated and comprehensive national strategy.

Implicit in his thinking from the beginning was the need to engage the public actively in the process. He had been deeply impressed while serving as chairman of President Reagan's AIDS commission by the public hearing process. At some early, tumultuous meetings on AIDS as many as 1,000 people were in attendance. Gradually the AIDS audiences settled down to more manageable numbers and the tone of the discussions moderated. A disparate group of commissioners under Watkins's steady leadership finally produced a remarkable degree of consensus on a national approach to this highly emotional issue. He sought now to model the energy hearings on the procedures used by the AIDs commission to gather information and gain public support.

Watkins believed that energy policy was like the AIDS issue in that the nation was divided regionally, economically, and ideologically. He felt that environmental groups in particular needed reassurance that a nuclear admiral was no nuclear zealot. In the case of energy strategy, he dispensed for the moment even with the mediating role of his own staff or of an advisory commission. The public itself would in a sense become the advisers, speaking its mind directly. Eventually he hoped some kind of broad consensus on an energy strategy would begin to merge.

The first phase of Watkins's tenure as secretary thus consisted of a kind of direct democracy. The public spoke, DOE listened. Although public hearings have been a part of American administrative practice for many years, and the use of hearings rose substantially in the 1970s, especially in state and local government (most notably, in public utility commissions, coastal commissions, and land use proceedings),[13] Watkins carried the public hearing concept to a new level. He made them a centerpiece of the policy process.

The initial five public hearings held around the country were almost wholly unstructured. The agenda consisted of a very broad question: how should the DOE allocate its funds for energy R&D? Either Watkins himself or Deputy Secretary of Energy Hinson Moore would chair the hearing. In all, 379 witnesses appeared at the hearings to exchange views with the secretary, deputy secretary, and later the deputy under secretary and from time to time with various other cabinet and subcabinet officers who chaired meetings at Watkins's urging. Written statements came from another 1,000 individuals and organizations, apart from the formal witnesses.

The role of the DOE staff was limited largely to recording the public comments. After the initial round of general hearings and after considerable criticism from various sources, including the OMB and his own bureaucracy, Watkins modified the approach and focused the sessions around more specific themes (for example, "Energy Services for Agriculture," "Public Health Issues in Energy Use," "Energy Use and Transportation Needs," "Environmental Implications of Fossil Fuels"). Thirteen additional hearings were held around the country in the fall of 1989 and early 1990.

In April 1990 the *Interim Report on a National Energy Strategy* was submitted to Congress and released to the public. The report was an unusual document by government report standards. It consisted of 230 pages of suggestions originating from the public hearings; all ideas were required to have a documented origin in one of the hearings or in the written submissions. There were "publicly identified goals," "publicly identified obstacles," and "publicly identified options" organized around the broad themes of (1) efficiency in energy use, (2) energy supply alternatives, (3) energy and the environment, and (4) the foundation of science, education, and technology behind the energy sector. The editorial role of the DOE staff was kept to a minimum. This compilation of ideas did not, however, add up to a coherent set of ideas, much less a national strategy.

The DOE staff that participated in the process told me: "We got a better idea of what was out there"; "Certainly the regional differences were clear. . . . It was not so much that we learned anything new, but we learned broad principles and got a sense of contentious issues";

"The sharp edges of these [contentious] issues were beginning to be rounded."

Many who were not a part of the process viewed it as an abdication of administrative responsibility, a year spent in rediscovering the basic facts of life. It left the agency, in their view, no closer to the trade-offs, the hard choices, and the careful analysis that would produce a coherent national strategy. The basic problems remained: the fragmented congressional jurisdictions and the squabbling constituencies blocking comprehensive action. The quest for a national strategy would continue to prove difficult and could not in any event emerge from a process that simply mirrored the divisions in public opinion.

The secretary's problems with his department continued to mount while he persisted in his populist stance. He made caustic references to the press about the narrow mentality of the DOE staff. They reciprocated with the time-honored Washington guerilla warfare against their boss, including leaks, end runs to Congress, and delay. At one point the story was spread that the secretary's famous temper had exploded during a frustrating telephone conversation with a DOE staffer, and he ripped the phone from the wall.

In any case, by the spring of 1990 Watkins saw the need for a new approach. It was in this context that he turned to the new SEAB for help. He hoped that SEAB would bring a more expert orientation to the planning process and would mold a new consensus approach. This change of direction marked a shift, in effect, from the populist to the technocratic Watkins. SEAB would help to be a bridge between popular aspirations and the complex technical dimensions of energy policy.

The first official SEAB meeting was held on April 22, 1990. The main agenda item was to review the interim energy report and to recommend next steps. Expertise manifestly had to be added to the processes of direct democracy in order to produce useful policy. And finally, wider use of internal DOE staff resources also became a part of the policy process as the secretary called a truce in the war with his own bureaucracy. Through elaborate effort over the course of the year and into 1991, the administration produced a final energy report that it sent to Congress in the spring of 1991.

Congress meanwhile had moved off on its own, with Senator J. Bennett Johnson of Louisiana, Energy Committee chairman, and ranking Republican Senator Malcolm Wallop of Wyoming taking the lead. The Bush administration supported S. 1220, the Johnson-Wallop bill, which contained strong production incentives for new energy development, permitted oil drilling off the shores of Alaska, and included a proposal to increase the corporate average fuel efficiency (CAFE) standard for automobiles.

The measure ran afoul of two groups of senators: environmentalists, who objected to the drilling provision on previously protected areas, and automobile supporters, who opposed the increase in the CAFE standards. S. 1220 was finally defeated on a cloture vote that its backers sought to bring the measure to a floor vote. The congressional-administration coalition needed sixty votes to produce a floor vote but fell short despite one of the most intensive lobbying efforts in recent years. Senate Majority Leader George Mitchell pledged to bring the bill up early in the next session, but Johnson, Wallop, and the administration forces led by Watkins resigned themselves to dropping the off-shore drilling and CAFE provisions. The national energy strategy was in limbo as of late 1991, with the affected interests, and especially the embattled American auto industry, preferring no bill to a bad one.

Where SEAB would go in the future could not easily be foretold. The secretary's environmental leanings seemed likely to be reflected in the board's evolution. The uncertainties and fluidity of the situation as of late 1991, however, had to be kept in mind. For a lengthy period Secretary Watkins had operated with a small team of trusted aides while only gradually filling the senior management positions within the department. He had yet to work out fully the interactions and staff relationships within his department. It appeared that significant changes could occur on short notice. The advisory system's impact on energy decisionmaking and its working relationship with the major operating divisions could be subject to significant change. The whole apparatus continued to bear the stamp of Watkins's unique blend of populist and technocratic leanings.

Conclusions

Watkins's DOE did not use its advisory committee in a typical manner, as a means of building a supportive constituency behind the agency's plans or to devise the plans that the agency would embrace. Such an approach implied too passive a role for the citizen, almost an insult to democracy. The DOE approach was to listen first to what private citizens wanted and only then to the judgment of experts (who were also outside the agency). After that the DOE was to derive policy from the preferences set forth by the people and the science advisers.

As policy developed, it became apparent that the public could not be a political actor speaking with a unitary voice to guide government action. The specialized publics, and especially those claiming to speak on behalf of the public interest, presented as always their own partial visions as the common good. The clash of values in the past had produced the DOE's program elements and now constrained the possibilities of easily dismantling those program clusters in the interest of some new vision. SEAB at least could, it was hoped, give Secretary Watkins and his team of close associates some pragmatic reform ideas and concepts to reorganize the DOE bureaucracy.

But as the DOE entered into interdepartmental negotiations and budget battles with the OMB in the summer and fall of 1990, energy policy was constantly forced back into the familiar patterns that existed before the public hearings began. As the action shifted to Congress in the following year, the powerful specialized constituencies and their congressional allies dominated the process and produced a conventional energy measure stressing production incentives. Blocked by environmental and other disputes at the end of 1991, legislation could be adopted in modified form in the next session of Congress, capping a three-year struggle over a national energy strategy, in the face of stubborn recessionary tendencies in the economy and the pre-election maneuverings of the political parties.

Although political realities have limited what could be accomplished by reaching outside the established framework, the public hearing strategy cannot be dismissed as quixotic. Administrative history is replete with attempts by top leadership to circumvent entrenched interests. Such efforts frequently fail. Admiral Watkins could hardly have succeeded in transforming his department through public hearings alone or through the aid of outside experts reporting to the top. Organizational change is a more demanding task. But Watkins's public hearing process was an ambitious effort to instill a new spirit and alter the style of a department beset with problems. Watkins sought to use the leverage of outside advice, expert and lay, to produce change in his department. In essence he tried to bring a more inclusive view of policy to a badly fragmented department and to educate the public on realistic benefits from technical advances in the energy field. In doing so he exhibited the role that the advisory committee can play in an overall strategy of change. Whether he succeeded in altering the bureaucratic culture will perhaps only be evident several years hence. He has had more longevity in the post than most previous secretaries, which should help him to bring his ideas to fruition.

In the future a major challenge that Watkins and his circle of advisers will face is to communicate their ideas effectively to the working levels of the organization. The expertise within the department will ultimately have to be tapped more fully. In the meantime Watkins has proceeded to use the new SEAB to clarify his own thinking on major problems and to help rethink departmental priorities as well as to educate the broader public.

Extensive interaction between the DOE leadership and SEAB members has produced four priorities for future SEAB effort: to examine the future of the national laboratories administered by the DOE; to devise a strategy and recommend steps to solve the problems of civilian nuclear waste disposal; to assist the secretary in planning for a new nuclear production complex for the nation's defense needs into the twenty-first century; and to develop expertise on economic modeling as related to energy costs, supply, and use. This list is clearly also a statement of some of the central problems facing the department and leaves little doubt that SEAB will be at least engaged in relevant work, whatever its ultimate impact on the department.

The secretary has also used outside science advisers in exploring the more narrowly technical issues that exist in numerous DOE mission areas. An illustration is the committee created in March 1990 to look into and advise on the future of the fusion energy program. In September 1990 the Technical Panel on Magnetic Fusion, constituted as a panel of the full board, issued its final report on the future of magnetic and inertial fusion energy development.[14] The report was a detailed technical review of the fusion program and recommended steps for pursuing both inertial and magnetic fusion from the research stage through demonstration project. This was to occur in the year 2025, with a view toward commercial operation in 2040.

The failings of the DOE's advisory system in the past related to larger matters over which the advisers and their patrons had little control. The ideological cross currents that largely immobilized the political system after the deregulation of oil and gas, combined with the low world oil prices, produced a situation in which budgets for many ambitious energy R&D projects were whittled down. The congressional committees that watched over their prerogatives made sure that no advisory effort would disturb the policy equilibrium. In this context the science advisers had nowhere to turn; the advisers cannot alter a political gridlock. They can only mirror it. Whether the new SEAB can function more effectively than its predecessor committee will depend on the interplay of the broad political forces shaping energy policy and on Secretary Watkins's ability to forge a modus operandi with his own department.

6 ||| Symbolic Politics and the Search for Public Support: The NASA Advisory Council

THE NASA ADVISORY COUNCIL was created in 1978 as part of a reorganization to comply with Carter-era pressure by the Office of Management and Budget to reduce the number of advisory committees (see chapter 2). The National Aeronautics and Space Administration amalgamated several previous advisory entities into the new council. The advisory council played little or no significant role in NASA decisionmaking in the first five years of its existence and only a marginal role thereafter. A remark of a senior NASA official to me is representative: "I can't recall a single instance in ten years where I ever heard the [NASA] administrator say, 'Hey, let's get the view of the advisory council on this one,' or 'I wonder what the advisory council thinks about this.' "

The advisory council has primarily performed the function of a support group. It has provided an opportunity for a body of distinguished citizens to be closely informed on NASA activities and to be public spokesmen on the agency's behalf. The NASA experience in this respect is not unusual. Many advisory committees do not actually advise their agency; support of the agency's mission, evidently, is some of what all advisory committees do and all of what some committees do.[1]

Although the NASA Advisory Council has been primarily a support group, this has not been the council's only role. Like the Defense Science Board and the EPA Science Advisory Board, the NASA Advisory Council consists of an overall board plus a number of ad hoc task force groups convened from time to time. In addition, there are seven standing committees (see figure 6-1), some with their own subcommittees, and one congressionally mandated advisory committee (the committee on safety, created in the wake of the January 1986

Figure 6-1. **NASA Advisory Council and Committees**

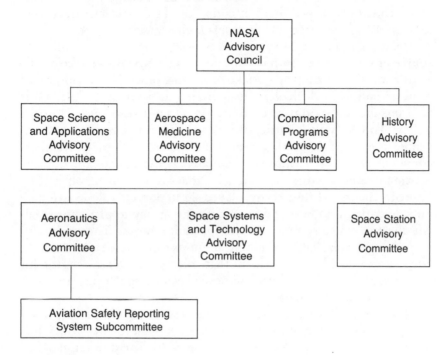

Challenger disaster). These panels or subgroups do engage in more serious and detailed analysis of problems and technical options facing the agency. In addition, the council has not remained totally frozen in its original boosterism mode. It has attempted at various points to channel ideas back to the agency and has on several occasions played a modestly useful role in bringing about improved coordination between NASA and the Defense Department.

A shift in the council's passive and traditional role began to occur under the 1983–88 chairmanship of Daniel J. Fink. Fink, a former General Electric Space Systems executive and Defense Department official with an action orientation, sought to make the work of the council more relevant and useful to the agency. As he told me in an interview, "When I became chairman, I discovered that what the council was doing had almost no relationship to the problems and issues that NASA was actually worrying about. I decided to try to focus our efforts on the real problems."

He attempted to accomplish this by critically examining policy assumptions rather than by simply being a cheerleader. He succeeded most notably perhaps in the study of the mixed fleet of launch vehicles.[2] This study helped focus attention on the deficiencies of relying exclusively on the space shuttle as a launch vehicle.

The work of the advisory council panel was not of course the only source of such advice. Both inside and outside NASA, reliance on the shuttle was seen as a policy mistake of the first order. Yet it had proved to be exceptionally difficult to even discuss the issue openly, let alone to bring about a serious re-examination and potential reversal of the policy. To justify the shuttle, more and more missions had to be invented for it and more flights projected to make the cost numbers come out right. The claims were inflated as the doubts grew, a distressingly familiar story with policies that are fundamentally wrong headed. The mixed fleet study played an important role in bringing outside criticism to bear and in forcing the agency to plan more realistically for a diverse range of transportation needs. The study also helped to galvanize DOD thinking about its own space transportation effort after it had become evident that sole reliance on the shuttle for military lift capability, which had never been a popular position with the Air Force, was untenable.

Nevertheless, the internal debate triggered by the mixed fleet study as it proceeded still might never have been enough to alter policy had it not been for the *Challenger* disaster in 1986. The disaster shocked the system enough to force a reappraisal of the space program, including the transport issues. The advisory council's efforts made some plans and alternatives available at the time the agency was receptive to and in need of ideas. The formal report on the mixed fleet, when it came out in 1987, reflected what were by then accepted facts: the shuttle alone could not be the launch vehicle for the whole of the nation's space effort; serious planning for alternative vehicles had to be undertaken. A successor project in which the council looked at long-run transportation needs of NASA and worked closely with a task force group from the Defense Science Board provided NASA with useful advice and has had a continuing impact on NASA operational plans.[3]

The utility of the advisory committee as the means to convene more focused efforts by a task force must be emphasized. Like the Defense Science Board, the NASA Advisory Council provides a mechanism by which talent can be assembled in a working group. The working group can be significant for the agency even if its parent body does little else. This facilitator function is perhaps the most important role that the overall advisory committee plays; it provides the framework within which focused efforts can take place.

Whether the working group accomplishes something depends on, among other factors, the timing of its report, the quality of the team assembled, and the ability of the group's chairperson to define a manageable task and form a close working relationship with agency officials. The task force can operate more informally than the parent body. The task force carrying out the mixed fleet study, for example, held a number of its early sessions on a "fact-finding" (that is, preliminary) basis. A fact-finding meeting does not need to be open to the public or announced in the *Federal Register*. But permission to hold a fact-finding meeting is normally obtained in advance from the NASA committee management official so as to conform to the agency's interpretation of the Federal Advisory Committee Act. A short report must be filed at the end of the year with the Library of Congress indicating the dates of all closed meetings and describing the topics discussed.

Any formal meeting of a council subcommittee or working group must be open to the public and announced in advance under NASA's interpretation of the FACA. Meetings of task forces or subgroups at the office level (that is, below the level of the six subcommittee or officially designated council task forces) are much more informal. The NASA Advisory Council, in brief, is the formal tip of an iceberg; a significant part of the activity takes place before formal meetings. But a NASA official, designated as the committee management officer, in theory supervises the whole process and sets guidelines for the component offices. In practice, committee management officials attempt to be helpful to their operating divisions and generally seek to facilitate what the divisions want to do.

NASA also has what are termed *operational working groups*. These groups have some attributes of an advisory committee but are deemed to be exempt from the FACA because they are *postdecisional*. That is, the group is engaged in a detailed technical assessment or planning function that is not intended to result in policy recommendations. The operational group may work closely with, or be a part of, a *utilized* advisory organization or consultant entity. An example of such an entity is the Synthesis Group. This group assembled information and options on the technical characteristics of four alternative space exploration strategies with the assistance of the Rand Corporation.[4] The Synthesis Group, headed by former astronaut and retired air force general Thomas Stafford, was created by executive order in 1990. Its purpose was to begin planning how to fulfill a goal articulated by President Bush on the twentieth anniversary of the lunar landing, a manned exploration of Mars in the next century. The report essentially laid out a strategic plan for the U.S. civil space program for the next three decades.

The seven formal advisory council standing committees that report at the level of associate administrator can be influential bodies in NASA operations and operational planning. They carry on and embody the traditions of in-house technical expertise and of outreach to the broader technical community, a style of operation dating from the old National Advisory Committee on Aeronautics (the predecessor entity set up in 1918 that grew into an operating organization in the 1930s and around which NASA was formed in 1958). The chairmen of the seven standing committees sit as ex officio members of the parent NASA Advisory Council.

The Space Science and Applications Advisory Committee (an amalgamation of what had previously been separate science and applications committees) provides an example of an active standing committee. It has played critical roles in helping NASA plan the evolution of its space sciences program, in lobbying for increased NASA space science budgets, and generally in mobilizing a supportive constituency behind space science. Among the standing committees it has the reputation as having the most strong-minded and independent members. The committee became influential under the chairmanship of Louis Lanzerotti of AT&T Bell Laboratories, a physicist with long experience in the space field and distinguished service on NASA committees. In 1988 he resigned as chairman to become the chairman of the National Academy of Sciences' Space Science Board. Lanzerotti's stature in the field was shown when he was also appointed a member of the Committee on the Future of the U.S. Space Program by President Bush in 1990.

The space science committee under Lanzerotti's chairmanship functioned as a kind of gatekeeper for major initiatives in NASA's space sciences program. It is doubtful that NASA would undertake any important new space sciences initiatives in the face of objections from the committee. The committee members are vigorous and influential spokesmen and -women who command respect in Congress and the media. Many testify regularly as authorities in their own right on space sciences issues. They do not testify as official spokesmen at budgetary hearings, but they are invariably supportive of the sciences part of NASA's budget.

The space science committee is the most active of the NASA standing committees, but the others play useful roles in evaluating NASA plans and in responding to specific requests for assistance. The ability of the standing committees to induce NASA to undertake programs is much more limited than their capacities to block or to force a reformulation of a proposed plan of action.

Beneath the standing committees is another layer of advisory activity at the divisional and office levels. These committees are very informal, and some parts of NASA do not have them. Nevertheless, the system is sufficiently structured to provide something of a career ladder in advisory service. Scientists get recognized and known through their service on an office committee or a divisional committee and may subsequently be appointed to one of the seven committees reporting to the associate administrators. The career of Louis Lanzerotti is illustrative. By dint of hard work on committees at the lower tiers, he was appointed to the space science committee and eventually became its chairman. As chairman he also became ex officio a member of the NASA Advisory Council.

A good example of an effective effort at the standing committee level is the 1986 report *Crisis in Space and Earth Science*.[5] This report, coming in the fall of the year of the *Challenger* disaster, took the accident as a point of departure to discuss a problem that had been developing for years. The problem, exacerbated by the *Challenger*, was the changing role of academic scientists in the heavily instrumented field of the atmospheric sciences.

Characterizing the report as a discussion of "big science" versus "little science" does not do it justice, but does indicate some of the issues explored. The report considered what portion of the budget should go to training, analysis of data, or small-scale projects instead of to large and expensive space payloads. It criticized the decision-making process by which NASA commitments are made, deplored the dependence of whole fields on a single space launch, and reviewed other problems intelligently and critically. The word *crisis* in the title suggests the tone and critical thrust of the report, although the report by no means resolved the issues it raised. It wavered at times in its analysis, sometimes seeming to endorse more small-scale efforts in universities. At other times the report appeared to limit the university role to the analysis of data, with little or no role in planning the big space payloads.

That the report could not resolve the issue is hardly surprising. The problem is extremely difficult. Yet the report posed a challenge to the NASA leadership and to the scientific community. An important cluster of issues needed to be addressed urgently, and the report focused attention on them in dramatic fashion. When the report attracted favorable publicity, it achieved several purposes: it called attention to the problem (and thus made it potentially more likely that congressional and media attention might move the agency toward action); it symbolized some degree of agency concern and focus on the problem; and it brought home to the universities that they needed to rethink the academic role in large-scale, space sciences projects. So long as the agency itself and the academic community had no clear answers to the questions raised, however, no immediate action was likely to be forthcoming. The consciousness raising function of a report like *The Crisis in Space and Earth Science* raises the question of inside influence versus outside advice and the leverage that comes from favorable (or unfavorable) publicity.

Insiders versus Outsiders

An interesting facet of the NASA advisory system is the role of the National Academy of Sciences in the system. The NAS Space Science Board and the NAS Aeronautics and Space Engineering Board have evolved a modus operandi with the internal NASA advisory bodies. The NAS committees largely steer away from short-term operational details, leaving these to the NASA in-house technical staff and advisory system, and focus on more long-range issues. In all there are some 800 to 900 boards and committees operating under the NAS complex, performing studies and reviews on contract for government agencies. The NAS advisory effort almost invariably results in a public report.

The NAS's original involvement with space sciences issues dates back to the Sputnik period when a space sciences committee was first organized to advise the president and Congress on what kinds of space experiments the United States should contemplate. This experience may have given the scientific community a slightly misleading model of the policy process. Because President Eisenhower in the Sputnik crisis responded to timely advice from the outside, the scientific community has frequently believed that if only it could get the facts directly to the president, the appropriate action would follow. The more normal pattern is for the president to be besieged with information of all kinds, and for each individual study or bit of advice to have a parochial or biased outlook and to be insufficient by itself as a basis for action. Studies generate more studies in a seemingly infinite regression. This process is not brought to cloture until some action-forcing mechanism or impulse from the policymaker finally cuts off the debate. The public studies that the NAS carries out, and its careful procedures, make its information and analysis almost always respected in Congress and the executive branch. But this information only rarely becomes the basis for clear-cut action by Congress or the executive branch.

Yet the NAS is routinely called upon by Congress for special assignments. And there are instances in which the system actually works to produce a clear result; for example, an NAS report effectively halted the planning for a small space station to be used for commercial purposes in advance of the larger *Space Station Freedom*.[6] In this case, the NAS was given a specific assignment by Congress, to evaluate the commercially developed space facility idea. It approached its task without preconception and came forcefully to the conclusion that the concept was not a sound one. As one member of the NAS panel said to me, "We approached the job seriously. No one was locked into any position at the start. But it soon became clear that [the proposal] just didn't make sense. We saved the nation billions of dollars. This was the most satisfying advisory effort that I've been part of."

Part of the logic behind the division of labor between the NAS and NASA sciences boards is evident: the "internal" external advisory board, in contrast to the "external" external advisory board (the NASA board vs. the NAS board), can be presumed to know more about the agency's current operational plans and programs, and its recommendations on specific issues presumably are more credible. The NAS body, presumably more detached, can more appropriately consider the longer-term issues. An important part of the strategy is also the self-conscious effort to gain maximum publicity and to leverage the influence of the different sets of advisers. Often of course one report is not enough to command attention; two seemingly different sources affirming the same policy thrust carry more weight. When one board feels blocked or considers that a dose of publicity might nudge the policy process in a desired direction or simply wants the prestige of

NAS endorsement, it passes along its views to its sister board for appropriate action.

The difficulty is that the target for the recommendations is often diffuse and the actions called for are often outside the agency's authority. If NASA can act on an idea recommended by the insider board, for example, there may be little need for resort to the NAS board. When NASA requires publicity to be pressured into action (and NASA itself may make this judgment), the reason is that there are obstacles to the proposed action. Often the barriers are fiscal, the high price tag for NASA's space projects. Publicity then becomes an effort to raise public awareness, build support behind an idea, and seek to induce the president, the Office of Management and Budget, and Congress to endorse the idea. It is easier to get endorsement for vague, far off goals than to get appropriations now. The whole effort becomes a shadowy game of affirming in principle a commitment to do something later and study the issue further. The result is that the airwaves are full of goals and debate but nothing happens. The politics of NASA advice giving is quintessentially the struggle for public support in the hope that a favorable climate will make a decision to commit resources more likely.

But the struggle can also be over stopping a program. It is doubtful whether the internal NASA advisory system could have been quite so bold and unambiguous in its critique of the small commercial space station proposal as the NAS panel was. Of course it is easier to stop a proposal though a critical study than to initiate a program through a favorable study. In this respect the advisory committee system is merely another manifestation of the wider political culture of checks and balances; it becomes another veto group and hurdle that must be cleared.

The advisory system thus reflects both the strengths and weaknesses of the pluralism of America's governing system. Sometimes ideas can work their way through the chain of committees until they become ripe for public release or for agency action. Or an internal dissent may reappear in another form and in another forum. The disadvantage is that the multiplicity of reviews and analyses can befog the issues; the blizzard of reports can create a kind of noise in the policy process. It is perhaps not surprising that multifaceted advisory structures reflect the pluralism of the policy process, but they can also magnify the disorderliness and confusion. Consensus can sometimes be harder to achieve, especially if there are charges by one side that the other is ignoring scientific truths—a claim that is difficult for the public and the political process to resolve.

A Lack of Dissent

In NASA, however, perhaps the surprising feature is how little dissent has come from its science advisers. In the entire NASA advisory structure, few critical ideas have typically come to the surface. Perhaps because the system is engaged so heavily in the game of boosterism and building public support, the public recommendations have been orchestrated along narrow lines. Unorthodox ideas seem to get lost, and advisory committees rarely break ranks with the established policy framework. For years no official committee raised objections to or challenged the reliance on the shuttle as the nation's exclusive space launch vehicle or questioned the manned space emphasis of NASA programs. Until the mixed fleet study the shuttle concept was never even seriously examined. Part of the reason was that the space flight side of NASA did not have the network of outside advisers that existed in other parts of the agency. The space launch bureaucracy insulated itself from critics and took refuge in remarkably unrealistic economic calculations purporting to show that the shuttle was cost effective. The NAS study referred to above, which effectively quashed the small commercial space station, was, however, enthusiastically received by space advocates because they thought it safeguarded the future of *Space Station Freedom*.

Numerous study commissions have been convened to look at NASA's future. But most have failed to probe deeply into the agency's problems. The 1987 Sally Ride commission illustrates the dilemma of the advisory group constrained from the start.[7] Given a narrow charter, it was effectively blocked from addressing the most fundamental issues facing the agency (the mission of the space station, the man in space emphasis, space transport issues, and priorities and choices among program areas). The commission ended up by endorsing a bold new concept "Mission to Planet Earth," a proposal to launch satellites to observe environmental trends, compile a data base, and guide national policymakers on environmental dangers. This was to be a new mission for NASA, but no suggestion was made on what to cut to make room for it.

Genuine dissent in NASA advisory efforts, in short, has been remarkably scarce. A notable exception is the work of the Advisory Committee on the Future of the U.S. Space Program, convened in the summer of 1990 at a critical juncture in NASA's history. This is a story worth telling in some detail.

The Advisory Committee on the Future of the U.S. Space Program

The advisory groups that were assembled to help plan for the development of intercontinental ballistic missiles in the 1950s or the development of a civilian space program in the 1960s knew that the nation had embarked or was about to embark on major new undertakings. The more recent space goals set by advisers have been of a different order, without timetables or resources for their achievement. The nation might or might not choose to do some of the things that were being proposed for the distant future. The uncertain fiscal climate cast an aura of unreality over the more ambitious projects. The members of the NASA Advisory Council knew this but felt locked into the pattern of seeking commitments in principle that might potentially lead to programs somewhere off in the future.

Everybody connected with NASA understood that in the climate of fiscal austerity NASA's fate rested in the hands of officials outside of NASA itself. The president and Congress would ultimately make the critical budget decisions. Yet NASA had done reasonably well in budget terms during the Reagan years and even into the Bush presidency. Reagan was a space enthusiast and liked virtually every big science project proposed to him.[8] Despite NASA's failure to get firm decisions on its most ambitious goals, many in NASA considered that the agency had done reasonably well in tough budgetary times. Some even felt that NASA could become the beneficiary of defense cuts, and strengthen its position as a preserver of the nation's technology base. A more sober assessment was that NASA's technology was old, its own infrastructure badly in need of modernization, its sense of mission confused, and its goals spread over a wide program front with too many starts and not enough resources to do any job well. A sense of malaise had in fact settled over the agency even amid the official boosterism that marked its public face. The problem seemed to be one of getting high-level attention to the agency's problems. Despite the rhetorical support it apparently enjoyed, NASA could not continue in a state of limbo.

In 1988 Congress created the National Space Council in the Executive Office, to be chaired by the vice president, as a partial answer to the agency's problems. Representative Bill Nelson, Democrat of Florida, whose district included the Kennedy Space Center and who had flown on a space mission, was instrumental in the council's creation. Soon after its creation the National Space Council began pressing NASA for a comprehensive plan for the future. NASA in response undertook a ninety-day internal study that reviewed broadly its major missions, goals, and program areas. The space council convened a blue ribbon panel of advisers to review the NASA plan. The experts found the plan seriously wanting and recommended the creation of a NAS committee to restudy the issues covered by NASA.

The NAS committee, chaired by former presidential science adviser H. Guyford Stever, subsequently came out with a report that was diplomatically phrased but clearly critical of NASA's approach.[9] The space council pushed for a new high-level advisory committee to look closely at NASA. The agency resisted strenuously. The stage was set for a critical look at key NASA programs—the space station design, the reliance on the shuttle, manned space flight, space exploration—when the agency's troubles multiplied in 1990.

A series of management failures and apparent blunders finally produced a crisis atmosphere, including the grounding of shuttle launches because of fuel leaks, the failure of the *Hubble* telescope to function as expected, and new findings suggesting that the repair hours needed for *Space Station Freedom* had been seriously underestimated. It seemed evident to close observers of U.S. space policy that something significant had to be done. In this context, the space council's push for the creation of an independent, blue ribbon advisory committee to review the future of the U.S. space program won approval. NASA finally relented when persuaded that it had no choice and that the visibility of the new advisory committee might bring the resources the agency needed to plan seriously for the future. The NASA Advisory Council was not seriously considered to perform the study and was not even officially consulted in the decision, but two members of the council were appointed to the new committee.[10] The new Advisory Committee on the Future of the U.S. Space Program, known popularly as the Augustine committee after its chairman, Norman Augustine, was deemed necessary because it could attract a prestigious membership for the short-term assignment, command visibility and public attention, and take a fresh and independent look at NASA's problems.

The Augustine committee gained a critical victory before it even began its deliberations. With space council backing, it was able to escape the limitations on its mandate that had crippled several previous committees. All issues were to be on the table. Augustine and his colleagues were to look broadly at the future.

The committee began its work in the summer of 1990 with a 120-day, self-imposed deadline and actually completed its work ahead of schedule on December 17, 1990. The committee held external public hearings with some 300 witnesses and also interviewed many NASA scientists and administrative officials at the major laboratories. The committee was an extremely serious, hard-working group that gave generously of its time, experience, and talent. The group had the elements for success: a sense of national urgency, an experienced and effective chairman, a blend of skills that complemented one another, and a receptive audience.

The report, which received wide and highly favorable publicity, was a model of outside analysis helping to break an internal policy gridlock. The committee proceeded on the assumption that the aim was to produce a consensus behind the space program. Many groups supported some kind of space effort, but no two groups seemingly supported the *same* space program. The committee produced an integrated program design with a rationale that made broad sense. The report said the overall goals of the civil space program should be to advance space science and space exploration, not to commercialize space. Specific programs could then be evaluated in terms of their contribution to the overall goals. The space station design was severely criticized as attempting to serve too many purposes and not effectively advancing any single one. A major redesign was called for to focus on the station's role in space exploration and space science, with the commercial microgravity manufacture project eliminated. The shuttle was criticized; shuttle use was to be limited in the future missions and eventually phased out.

Without directly joining or fully resolving the issue, the committee moved NASA toward a more limited role for the manned space program. Manned crews were explicitly not to be used for missions involving only the placement of commercial satellites into orbit. Much of NASA's near-earth effort (its Mission to Planet Earth) was to be directed toward advancing knowledge of the environment, and the advancement of science was to be an overriding goal generally for NASA activities. The committee also made recommendations on management reforms for complex federal technical programs, citing the inadequate management structure for the Mission to Planet Earth project office. Space exploration was to be on a "go-as-you-pay" basis. The committee urged steady 10 percent real budget increases for NASA over the coming decade to achieve the goals set forth.

After submitting its report in December 1990, the committee officially disbanded but left open the prospect of convening again to take stock of the situation. The NASA Advisory Council was given the role of assisting NASA in implementing the recommendations. NASA declared its broad agreement with the Augustine committee's recommendations, but it was evident that the policy battles over the space station and earth orbiting laboratory redesign, the future of the shuttle, man in space, and other issues were by no means fully resolved. The Augustine committee had nonetheless served NASA and the nation well in sharpening the public debate and finally joining the principal issues. The interrelations between the military and civilian space programs and the role space could play in promoting the development of commercial technology remained major issues requiring the efforts of some new advisory panel in the future. In an effort to avoid the absurdities of the spin-off arguments of a decade earlier, Augustine and his colleagues eschewed claims that the space program would pay off handsomely in civil sector technology development. But they perhaps missed a chance to identify space with the popular cause of competitiveness and arguably could have urged an alternative rationale— that is, civil technology development rather than the promotion of science as the central rationale for space activity.

The most critical part of the Augustine committee's report was the call for a 10 percent annual increase in NASA's budget for the rest of the decade. It is too early to tell how this recommendation will fare. Congress's handling of the space appropriation for fiscal 1992, however, does not give high confidence that it is willing to vote funds of this magnitude. The scientific community meanwhile has broken ranks and publicly quarreled over the space station concept. In the past scientists usually quietly acquiesced in such mega-projects as *Space Station Freedom*, hoping that basic research would benefit indirectly from the large-scale technological ventures. Scientists representing a number of professional associates were now criticizing the space station, using such terms as "orbiting pork barrel."[11] This split in the scientific community will undoubtedly affect and complicate the task of building a consensus behind space goals. The scientists assaulting the space station cast their argument in utopian rationalist terms: "There is no scientific justification of a permanently manned space station. That message was carried by Nobel laureates in Congressional testimony, major scientific societies in uncharacteristically blunt public statements, and by individual scientists in articles and letters and visits to congressional offices. The consensus was overwhelming."[12]

The critics assumed that whether or not the nation should build the space station was a scientific question. They confused the debate and weakened their case by this line of argument. Landing a man on the moon, orbiting an earth-observing satellite, or pursuing research on aeronautics—none of these past or present NASA missions was a purely scientific decision. Nor is there a fixed budget for research and development that would allocate any funds denied to large-scale technology programs directly to science projects. One result of the assault on the space station appropriation is that the numerous lines of fracture in the scientific community will become more visible to the public. The battles over resources for the many subgroups within science may become more acrimonious, and the authority of scientists will be correspondingly diminished.

7 ||| Strangers in a Strange Land: Science Advisers in the State Department

THE STATE DEPARTMENT has been by virtually any measure the least successful among the agencies studied for this book in blending scientists and science advisers into policy deliberations. The increasing postwar importance of science and technology for diplomacy has given rise to repeated and continuing efforts to integrate science advice into the foreign policy machinery. Many able officials have tried to deal with what they perceived as shortcomings in the department's handling of complex technical issues or emerging needs in a technologically more sophisticated commercial and political world. Scientists and diplomats have tried hard to work together, but for the most part they haven't succeeded. This chapter tries to puzzle out why this is so. Some very simple factors are partial explanations: the department has not allocated time, money, staff, or psychic energy to the task; and the scientists have been impatient in learning the nuances and complexities of the agency they have sought to advise and have exaggerated their own importance. Beyond such factors, however, lies a puzzle over why so many efforts have met with such limited success. By seeing failure clearly, one can perhaps learn more than by celebrating success.

Background

The story has a convenient beginning in the aftermath of World War II. The State Department's May 1949 Reorganization Task Force No. 2 observed:

The Department is dealing on the one hand with foreign policy matters which have a great effect upon United States scientific pol-

icy and on the other hand with international scientific activities which
have an impact on foreign policy. These matters are being handled
at various points without adequate scientific evaluation. . . . We
believe that the extent of the Department's responsibility for international scientific matters requires top policy consideration and the
aid of professional scientific judgment.[1]

In 1950 a State Department committee on science and foreign relations chaired by Lloyd V. Berkner concluded that "present [State De-
partment] organization is inadequate to assess with accuracy the nature
of the broad policy issues involving science."[2] It recommended the
creation of a unit within the Department of State responsible for the
scientific aspects of diplomacy. Thereafter a small office was established to administer a program of scientific exchanges. The office was
also to assist the National Science Foundation, the Department of Agriculture, and later NASA in their international programs.

The office grew in importance especially after Herman Pollack, an
able and energetic foreign service officer, became its director in 1964.
In 1967 the status of the office was raised when Secretary of State
Dean Rusk expanded its staff and responsibilities. It continued to operate, however, somewhat on the fringes of mainstream State Department activities. The ambiguity reflected in the Berkner report about
whether the "role of science and technology in foreign policy" meant
policy *for* science or science *in* policy, or both, left the office's mandate unclear.

The "cardinal principles" on which the Berkner report based its
recommendations seemed to many observers contentious or merely
hortatory: "United States foreign relations with respect to science must
take on a more positive and active character. . . . Closer relations between the Department of State and United States science must be established." What did these phrases mean in practical terms? Could the
United States actually make fuller use of the nation's scientific resources to advance foreign policy aims? Or should scientific exchanges be considered merely an aspect of cultural diplomacy? In the
latter case science would be an important but lesser aspect of foreign
affairs.

The real business of diplomacy for promising young foreign service
officers remained the traditional bilateral negotiations conducted primarily by the department's regional bureaus. The science office, in
short, was not a choice assignment for the aspiring foreign service

careerist, and the office thus had difficulty recruiting. The blend of skills sought was also a vexing question. In particular it was difficult to know whether to recruit outside scientists for temporary periods or build a scientific career track within the foreign service. Pollack labored on throughout this period. He continued to enjoy a modest success, largely because he had strong backing from Secretary Rusk.

Early in the Nixon administration, Pollack persuaded Secretary of State William Rogers to create a science advisory committee for the department, which would report directly to Rogers. The idea was to supplement the skills of the science office and to reach out generally to the wider scientific community. Members of the committee included former secretary of state Dean Rusk, Simon Ramo of TRW, Eugene Skolnikoff of the Massachusetts Institute of Technology (a former State Department officer), and others. The committee was organized into separate work groups and took on various assignments relating to space, nonproliferation issues, marine sciences, and scientific exchanges. Rogers met with the committee occasionally, and the committee enjoyed a modest success in expanding the resources and planning horizon of the science affairs office.

In 1973, Democratic Senator Claiborne Pell of Rhode Island and Republican Senator Howard Baker of Tennessee pushed proposals in Congress to create, respectively, a Bureau of Oceans and a Bureau of International Environmental Affairs within the State Department. These two proposals were combined into a Bureau of Oceans, International Environmental and Scientific Affairs (OES) during hearings on the State Department Appropriations for fiscal 1974. Congress in establishing this new bureau also stipulated that it should be headed by an official with the rank of assistant secretary. The new bureau incorporated the existing science affairs unit.

The bureau formally came into existence in October 1974, the only State Department bureau created by statute. The principal program emphases were on environmental regulation and the control of technology generally, reflecting the prevailing assumptions in domestic science policy at the time.

In the meantime, however, with national security adviser Henry Kissinger succeeding William Rogers as secretary of state, the Science Advisory Committee fell into disuse. Kissinger chose to rely on informal advisers rather than to convene the advisory committee; in particular, he relied heavily for technical advice on his former Harvard colleague Paul Doty, a chemist with broad experience in arms control issues. In Kissinger's view, events were simply too fast-paced to make an advisory committee useful in the diplomatic context.[3]

Former professors or scientists with a well-developed regard for their own intellectual capacities have been among those least interested in advisory committees. When Dixie Lee Ray, a scientist and a former commissioner of the Atomic Energy Commission, became assistant secretary of state for the newly established bureau in January 1975, she abolished the Science Advisory Committee.

The OES was actively involved throughout this period in a variety of intergovernmental science agreements, interagency consultations, negotiations, licensing and enforcement actions, and in a variety of environmental, ocean, population, and other matters. The bureau played an important role in negotiating the U.S.–USSR Agreement on Scientific Cooperation during the Nixon détente era, as well as bilateral agreements with a number of other nations. Another major area was fisheries and ocean issues (in particular, new responsibilities resulted from the Fishery Conservation and Management Act of 1976, Public Law 94-265; the National Advisory Committee on Oceans and Atmosphere Act of 1977, Public Law 95-63; and the call for international negotiations on the Law of the Sea, which the United Nations initiated in 1974). While not among the most powerful bureaus, the OES was active and enjoyed moderate respect among the department's senior officers.

The OES assumed major new responsibilities in the administration of the Nuclear Nonproliferation Act of 1978, Public Law 95-242. Also in 1978, the passage of the Foreign Relations Authorization Act for fiscal 1979, Public Law 95-426, section 504(a), required the OES to assume important coordinating responsibilities for major marine science and technology agreements, as well as a host of other international science and technology activities.

Title V of the Foreign Relations Authorization Act should be seen as perhaps the high point, at least in what was expected of the bureau. Title V assigned to the State Department and the bureau wide responsibilities for "coordinating" international science and technology activities and mandated the department to prepare an annual report on the international science and technology activities of all U.S. departments. The State Department and the OES were also mandated under Title V to create a planning staff. The purpose was to achieve greater policy coherence in the far-flung fields in which the bureau had operational or coordinating responsibilities.[4]

Assistant Secretary of State Thomas R. Pickering and I (who became the bureau's first planning deputy) decided to recreate a science advisory committee to assist in the task of planning and policy formulation.[5] The committee would, it was hoped, supply knowledgeable advice, bring a heightened awareness of science and technology to the department, and assist in the administration of Title V. The role would be much like the old committee's role under Herman Pollack, except that in this case the impetus for the idea came from a recruitment. As in the creation of the Energy Department's Energy Research Advisory Board (ERAB), the revised State Department committee was the result of negotiations between the policy-level official and a deputy he wished to recruit. Pickering, a career foreign service officer, was persuaded that he needed to reach out to the scientific community. Taking as his planning deputy a science policy scholar who would work with a prestigious advisory committee would help Pickering tap the views of the scientific community.

The policy portfolio of the OES at this time can be inferred from an informal study of the allocation of the assistant secretary's time among his different functions: 60 percent went to nuclear nonproliferation and export licenses of nuclear materials; 20 percent to oceans and fisheries issues; and the remainder was scattered among space, environmental, renewable energy, scientific cooperative agreements with other nations, and population issues.[6] Part of the task of the advisory committee would be to help plan for an effective integration of the disparate sides of the OES's work, to identify the crosscutting issues of wider importance to the department, and to anticipate future problems that did not fit within the established jurisdictional niches.

Re-creation of the Advisory Committee

Even though the bureau strongly wanted the advisory committee, the task of actually putting it together and getting the necessary approvals proved to be formidable, largely because of opposition by the Office of Management and Budget.[7]

The OMB had theoretically transferred its responsibilities for administering the FACA to the General Services Administration as of 1978. The GSA, however, had not yet determined what to do with the committee management function, and the small unit transferred from the OMB had drifted around the GSA with no clear organizational home. It was not yet sufficiently settled down to be able to conduct business effectively. Such at any rate was the conclusion reached by State Department officials. Even though the OMB had ostensibly shed the function, it continued to assert the right to review departmental proposals for new committees. The State Department formally submitted its request to create the committee in the spring of 1979, with charter, list of members, legal memoranda, and other accompanying documentation. It was submitted over the signature of Deputy Secretary of State Warren Christopher.

The OMB found "insufficient justification" for the creation of such a committee and denied the request (under the law, the department was required only to "consult" with the OMB). Though advised by counsel that the OMB could not block the committee, the OES decided that discretion was the better part of valor and chose not to fight openly with its budget examiners. Accordingly the proposal was revised and resubmitted with more elaborate justifications and supporting material. Again a midlevel OMB official turned down the request from the deputy secretary of state. An OMB official observed to me during a lull in the negotiations, "This committee would simply be used to dream up more ways for the State Department to spend money, and we don't want that."

The committee was finally created at the end of 1979 after the department decided that it had consulted sufficiently with the OMB. The president's science adviser played a role in backing the concept of creating the committee with senior OMB officials.

The laborious and lengthy process required to create the committee was a sign of the difficulties to come. Indeed the State Department committee is an example of the failure to develop a meaningful role even though the committee continues a perfunctory and pro forma existence.

In the previous case studies it was observed that no committee can spring full-blown into an influential role; the advisory committee has to struggle to blend its perspectives effectively with the agency's goals and mission. The culture of science and the agency's organizational culture may clash; and a lengthy period of mutual accommodation is required before the committee is able to render useful service. In the State Department's case, the inherent difficulties of blending technical advice with the nuances and fast-paced developments of diplomacy were formidable to begin with, and the failure to devote serious efforts to overcoming the difficulties doomed the effort almost from the start.

An initial minor difficulty grew out of the Ethics in Government Act of 1978. The department's first choice as chairman of the new advisory committee was Charles Robinson, a California businessman who had previously served as deputy secretary of state. He declined the invitation because of conflict-of-interest fears under the new legislation. The department's legal office had ruled that serving as chairman of the new committee would make him subject to financial disclosure requirements, but that he need make only short-form disclosures. Further the disclosures would be confidential and would be reviewed only by the secretary of state. A twenty-page, single-spaced legal memorandum accompanied the letter of invitation to Robinson. The lengthy legal opinion did not allay his fears (and may have heightened his uneasiness over the law's application). Robinson's decision was a disappointment to the OES officials planning for the new committee, who concluded after some further exploration that a number of other candidates had similar concerns. Assistant Secretary Pickering finally dealt with the problem of recruiting a chairman by deciding to act as chairman of the committee himself.

A second problem came with my departure as the OES planning deputy, since I had been the official most deeply involved with the committee. My successors did not have the same interest and commitment to the concept of outside science advice. The initial enthusiasm within the bureau for the committee had gradually waned as the problems dragged on.

Third, the committee's creation over OMB objections was a Pyrrhic victory in that the department decided it could not or should not allocate funds specifically for the committee's administration. Members were never compensated and did not even receive travel expenses.[8] It had been initially envisaged that the committee would have a secretariat, travel and expense money, and modest funds for contract studies. Without resources it could be safely assumed that the committee would not be able to do much.

The committee held its first meeting in January 1980. It generally met only twice a year thereafter, never for more than a full day and often for only a half day. Virtually no activity occurred between the formal meetings (with a few exceptions, to be noted later). The idea of creating subcommittees with specific areas of responsibility was discussed, but there was no follow-through. The idea was never carried into practice.

Explaining the Failure

The most critical problems ultimately lay in reconciling the culture of diplomacy and the scientific outlook, and in thinking through the committee's role. Did it have an assignment that mattered to anybody in the OES and the State Department? The traditions of open debate, peer discussions, and an analytic framework for decisions simply did not fit well with the traditions and incentives that have marked the careers of most foreign service professionals. Traditional bureaucratic rivalries within the department limited the advisory committee's scope and influence; the powerful regional bureaus continued to dominate the functional bureaus on most key issues. The advisory committee thus reported to a second-rank bureau, the OES, where it was not taken seriously, just as the OES itself was not taken seriously within the department (except on a few issues—notably, the proliferation of nuclear weapons—that were excluded from the committee's deliberations).

Virtually none of the bureau's operating officials had any incentive to work with the advisory committee. The turnover at the assistant secretary and the deputy assistant secretary levels made it difficult for the committee to develop close working relationships with the leadership. The small attentive public following the situation lamented the lack of progress toward making science a vital presence within the department, but it was in no position to effect significant changes.

The OES was unusual in being created by statute and also in having numerous statutory mandates. It was engaged in carrying out tasks mandated in at least fifty laws (for example, export licenses for nuclear fuels, attendance at international whaling conventions, fisheries negotiations, bilateral science and technology agreements, and annual reports required under various laws). These activities meant that bureau employees were extremely busy, operating under tight deadlines, in production-type activities. The advisory committee appeared to have little relevance to these tasks. A part-time committee could never learn enough in detail to be useful to the operating officials, and in some cases, such as the granting of licenses, it was inappropriate to have an outside group involved. The operators usually felt that it would be an unproductive drain of time to involve the advisory panel members in their activities. That the advisory committee was the brainchild of an outsider and that it was tied to the planning office did not help. In the State Department, as in business, separating the strategic planning function from the operating divisions did not work. Planning needs to be combined with operations to be effective. The catch-22 of the OES appeared to be that the operators were too busy to plan, and the planners lacked the detailed knowledge to formulate plans. To the operators it seemed to be a case of the planners who did not understand their job trying to advise them, and the advisers who understood even less advising the planners.

The broader policy aspects of the bureau's work, in particular the nuclear nonproliferation issue, were excluded because of secrecy considerations and because of the extreme sensitivity of the issues. Most nonproliferation issues were highly classified and involved the timing and content of high-level diplomatic overtures to heads of state. Pickering believed these were not matters that could be appropriately discussed at a public advisory committee meeting. The possibility of securing clearance for the committee members was rejected as impractical. Given the sensitivity of the diplomatic measures in progress, the prospects of preserving secrecy with a large group were not favorable. Nor could committee members be kept informed on a day-to-day basis anyway. Events simply moved too fast.

Pickering preferred to get technical advice from a consultant, Harvard professor Joseph S. Nye, Jr., who had served from 1977 to 1979 as under secretary of state in charge of nonproliferation policy. Nye was intimately familiar with U.S. nuclear policy and had a comfortable working relationship with Pickering. From the start the bureau's most important policy area was thus excluded from the province of the advisory committee.

The second most important policy area for the OES had customarily been oceans, marine science, and fisheries issues. These issues were removed from the committee's consideration by virtue of the strange practice in the State Department of appointing roving ambassadors who are superimposed on the line departments. Elliot Richardson had been appointed special ambassador in charge of the U.S. delegation to the Law of the Sea Convention negotiations. He made oceans policy, although the OES provided him with staff support.[9] Since the bureau leadership did not wish to preempt or appear to grab responsibility from Richardson, it resisted any effort to advise him or pressure him through the advisory committee. Richardson did not look to an advisory group for assistance on what he considered very delicate negotiations with foreign leaders. The absence of a base of public and congressional support behind the Law of the Sea Convention, which later doomed the negotiations, may be attributable in part to the failure to air the issues thoroughly with the OES advisory committee. When the advisory committee finally did get involved, in the Reagan administration, it was to limit damage and to enlist the committee's help in explaining to the public the administration's policy on the Law of the Sea issue. The bureau's two major issues were in any event not even discussed with the advisory committee during the first year, virtually ensuring that the committee would be peripheral.

It must also be noted that the electoral cycle played a part. The November 1980 election produced a new president and put policy on hold, especially given the criticisms of the Carter environmental, nuclear and nonnuclear energy, and Law of the Sea policies by the Reagan team during the campaign. These were central areas of OES concern. Although the advisory committee had been specifically nonpartisan in its composition, a Carter-era advisory committee was bound to cause concern with the constituencies that closely followed State Department activities in nuclear, environmental, and oceans policy. The transition was bound at least to cause delay while the new team considered its options.

The advisory committee's failure to achieve a significant role can finally be partly attributed to a long-standing and unresolved dispute over the appropriate staffing for the OES. This dispute persisted under the Carter administration and had its roots back in the bureau's early days. It was a problem irrespective of who was head of the OES. The OES had acquired a staff that was more technical than other State Department bureaus. Of the eighty professionals in the OES in 1979, forty-four had graduate degrees in technical areas and forty-four had had experience in mission agencies with extensive R&D activities prior to their OES service. This heavy technical emphasis contrasted with the more typical foreign service pattern of recruiting entry-level generalists. It was the result of earlier recruitment efforts based on the theory that the bureau needed more technical expertise. This approach was emphasized during the tenure of Dixie Lee Ray as assistant secretary.

After Dixie Lee Ray, however, senior foreign service officers felt that such expertise had not produced the desired results. In fact, the technical specialists often had had difficulty in adjusting to the State Department style of operation. The split between the technical and the foreign service orientations continued to plague the bureau and the department. One view held that the infusion of technical expertise was necessary to strengthen the department's ability to handle the increasing number and complexity of scientific questions. This was premised ultimately on the utopian rationalist conception of decisions as stemming inexorably from the facts. More technical expertise would lead, it was presumed, to sounder decisions. The alternative view maintained that the real need was to integrate a science and technology awareness into the culture of the foreign service. According to this view a foreign service officer should be taught enough of the technical substances of issues for him or her to function effectively. This was considered easier to accomplish than teaching a technical person how to operate within the complex State Department bureaucracy. Assistant secretaries after Dixie Lee Ray wavered back and forth, trying to achieve a synthesis of the two approaches.

At the time of Pickering's appointment as assistant secretary for OES in September 1978, the pendulum had swung to the view that the bureau needed full integration into the department. Further the bureau needed stronger central management. In keeping with the Title V mandate, the fragmented international science and technology activities of other departments should be more closely coordinated with the overall objectives of U.S. foreign policy. In the hierarchical style of foreign service operations, career officers did not freewheel in their activities. Indeed the tendency to set off on unauthorized initiatives was deemed to be one of the sins of the technical experts of the Dixie Lee Ray era. Senior and midlevel OES officials were thus preoccupied with trying to integrate themselves more fully into departmental operations. As a result the science advisory committee was born at the wrong time. The vague desire for outreach to the scientific community was not enough of a mandate or mission to create the committee. The OES was ambivalent about whether it really wanted the committee in the first place. The bureau and the department simply did not know what to do with the advisory committee. The committee members wondered what they were called upon to do and looked for guidance that did not come. Both sides saw the committee's creation as a modestly useful start, however, and looked forward to potential improvements as experience was gained.

The Reagan Era

No such improvement occurred. With Pickering's departure to become ambassador to El Salvador in February 1981 and with James

Malone's succession as assistant secretary, the bureau's fortunes turned downward. Malone unwisely pledged during his confirmation hearings that he would recuse himself from nuclear policy matters because he had once represented a nuclear power utility. In thus removing himself and the bureau from nuclear nonproliferation policy, he lost his most important responsibility. Nuclear policy was transferred to Richard Kennedy, a former commissioner of the Nuclear Regulatory Commission, who became a special ambassador in charge of nonproliferation issues. Malone decided instead to direct his energies to the Law of the Sea negotiations and other oceans issues. President Reagan's first secretary of state, Alexander Haig, had virtually no interest in the oceans or most other science and technology issues, and Malone was left to focus on extricating the United States from the Law of the Sea negotiations.

The OES pursued an assertive policy in reversing the U.S. position on the Law of the Sea. The proposed convention's commitment to the concept of the "common heritage of mankind" and to an independent international entity with taxing powers was anathema to conservative Republicans.[10]

Malone found a modest use for the science advisory committee in this context. Since the committee had prominent Republicans as well as Democrats, and had been indeed carefully chosen to have a bipartisan character, Malone and his colleagues decided that it was not a tainted Carter-era committee after all. Malone tried to use the committee as a forum for explaining the shift in policy on the Law of the Sea negotiations. He hoped that it could lend support to his and the administration's position, an effort in which he had some success. The advisory committee was thus primarily used as a sounding board. It was both a test of the feasibility of Malone's ideas and a support group to help persuade elite opinion that opposing the Law of the Sea Convention was justified.

No effort was made, however, to meet more frequently, to provide staff to the committee, or otherwise to make full use of it in bureau operations. While there had been a hope that the secretary of state would meet with the committee, it became evident that Secretary Haig had little interest in the bureau or the committee. There were few contacts with the highest levels of the department. The committee continued in its more or less comatose state, largely inactive but showing occasional flickers of life.

The fortunes of the OES improved with the appointment of John

Negroponte as assistant secretary for OES in July 1985. Negroponte, a senior career foreign service officer, had previously served for two years in the bureau as deputy assistant secretary in charge of oceans policy. He brought experience and energy to the bureau. He had served at the time the advisory committee was established and he believed in the committee's utility.

Negroponte's term witnessed the beginnings of a modest revival in the bureau's and the advisory committee's fortunes. He gave thought to how he might use the committee, and meetings actually had a focused agenda; usually one topic was discussed in depth. He knew and liked many of the committee members, found the meetings useful, and saw no purpose to be served by changing the committee's mandate or membership. He did not, however, significantly broaden the committee's responsibilities or devise important new uses for it.

The committee members, for their part, generally felt that they were continuing to render some useful service. They enjoyed keeping in touch with the bureau's problems and hoped that with time the position of science within the department would be strengthened. The membership remained virtually unchanged. The stalwart group who had fought for increased science and technology awareness in the State Department for many years stayed with the committee with a dogged determination and missionary zeal.

In 1988 Negroponte was succeeded as assistant secretary by Frederick Bernthal, a commissioner from the Nuclear Regulatory Commission and a close associate of Senator Howard Baker. Bernthal's extensive technical background and wide-ranging contacts in the scientific community brought a new, more outward-looking style to the OES. He felt comfortable with the science advisory committee and made a serious effort to integrate its work with bureau operations. Bernthal arranged for Secretary of State George Shultz to attend several meetings (which still, however, remained on a semiannual schedule). The committee became somewhat more active, with OES staff encouraged to make informal contacts with the advisers when they needed advice. The committee's stature inceased modestly. As one member told me with reference to the Negroponte-Bernthal period, "I think it [the committee] has gone from about a three to a four or maybe a four and one-half on a scale of ten in the past few years."

Perhaps the committee's most notable effort came in the area of international environmental concerns. Committee member Robert White, president of the National Academy of Engineering, led an effort to alert the department to the global warming issue. A report prepared by White, drawing on the resources of his organization, led to a study that had wide circulation within the Department of State. It was influential in alerting at least some high-level officials to the importance of this issue. A number of informal briefings took place between committee members and senior officials. An October 28, 1988, speech by Secretary of State Shultz, "The Ecology of International Change," delivered at the Commonwealth Club of California, drew on the study's analysis on greenhouse gases and global change. Earlier that year, the Montreal agreement on CO_2 was a notable achievement of OES negotiators, led by Ambassador Richard Benedict, deputy assistant secretary in charge of international environmental issues.

Conclusions

Despite the committee's increased activity in the mid-1980s, it still had not achieved a significant role in the scheme of things at the State Department by the end of the decade. The agendas for the meetings were rarely fashioned so as to focus on action items. The committee's concurrence or active involvement in current or evolving issues was rarely sought. The meetings tended to be mostly briefings on the bureau's recent work in various policy areas.

The committee had clearly come to report as a practical matter to the assistant secretary for OES rather than to the secretary of state. Secretary Shultz's occasional attendance at meetings enhanced the committee's prestige but did not lead to closer ties with the secretary's office. Secretary James Baker in the first years of his service took little interest in the committee's work.

Surprisingly, perhaps, the advisory committee had little turnover in membership over the decade of the 1980s. Members evidently have wanted to continue to serve. They often wondered what their role was as they reported for the semiannual meeting, but then six months would go by and they dutifully reported (most of them) for the next meeting. The succession of assistant secretaries found the committee somewhat useful or at any rate unobjectionable. At least they did not question the committee's existence, even if none of them formulated a mission for the group. The assistant secretaries apparently wished to avoid the criticism they would receive for discontinuing the committee and assumed that somewhere in the misty past there must have been a rationale for the committee, even if they could not say what it was.

The committee's position within the State Department may be illustrated by a short anecdote. At a June 1990 meeting in New York City that I attended, the following revealing episode took place. The meeting was part of a conference on the role of science and technology in foreign affairs sponsored by the Carnegie Commission on Science, Technology, and Government. The event in question was a dinner meeting launching the conference, with a former high-ranking State Department official giving the keynote address. Several members of the science advisory committee were in attendance. The keynote address was intended as something of an inspirational message, energizing the conference participants to tackle the problems of making technical advice a more integral part of State Department operations.

The speaker, warming to his subject, deplored the department's lack of expertise in science and technology. The ineptitude of the department was deplorable. It was a barren landscape drastically in need of change. He cited as a case study his struggles to inform himself on the acid rain issue prior to a U.S. visit by Canadian Prime Minister Brian Mulrooney. He criticized the department for its failure to develop adequate technical expertise. He spoke in the fashion of a revivalist condemning sin. Finally he explained that he took recourse to outside expertise in such circumstances, referring to some "twelve to twenty occasions" in which he arranged for informal briefings by groups of outside scientists. The message was clear: the State Department had to clean up its act.

This was too much for a member of the science advisory committee in the audience. Rising to lead off the question period and flushed with anger, the questioner lit into the speaker: "Weren't you aware that you had a science advisory committee in your own department? Why didn't you turn to *it* for advice?" The ensuing discussion, I think it is fair to say, did not fully reflect the upbeat spirit and inspirational message intended for the opening session. It did, however, bring out clearly the differences in perspective between the decisionmaking consumer of advice and the scientist-purveyor of advice. In chatting privately after the talk, the speaker clearly displayed his preference for informal discussions of the kind he arranged through Frank Press of the National Academy of Sciences. He observed, in terms reflective of the attitudes of many policymakers I interviewed for this study, "I hate formal committees. They're too bureaucratic. I went to considerable lengths to get the advantages a committee can offer without actually having to create one." Policymakers do not like to be constrained by formal structures. They will seek advice when, how, from whom, and as they think they need it, and will resist having anyone else define their advisory needs for them.

Curtis Bohlen, who became assistant secretary for OES in 1990, has evidently concluded that it is time to rethink the larger issues of science's role in foreign affairs as well as the missions and responsibilities of his bureau. He has decided to recharter the advisory committee with a fresh statement of mission, shorter terms for the members, and new operating procedures. He sought the advice of the committee members on how to refocus its efforts and to become more usefully engaged in the bureau's work.

At the first meeting of the committee since he took over, which was held in January 1991, Bohlen made the committee's role an agenda item. The initial message from the committee members was loud and clear: "Use us or lose us!" As one member present at the meeting recalled, "It was very unusual. You have to give him [Bohlen] credit. He listened patiently, and was very candid, very forthright. We went around the room and practically everyone dumped on the State Department, telling him why the committee wasn't effective and what was wrong with the department."

A brighter future for the committee perhaps lay ahead. At least serious thought was being given to the committee's role, composition, and organization. The bureau faced the challenge to think through how, in the fast-paced world of diplomacy, outside technical advice can be brought into the policy process. While there were no easy answers, it seemed clear that the use of outside advice must be accomplished in a way that conforms to the mores of this talented but bureaucratically complex department. At the same time it should aim to transform gradually the department's habits of thought and styles of operation.

The role of science and technology in the conduct of foreign affairs has paralleled the changing emphases in domestic science policy. The initial involvement of the State Department with science and technology was primarily to use America's leadership in basic science for the purposes of promoting U.S. prestige. The nation sought to support basic science abroad, just as we were fostering the advance of science at home. With the shift of social priorities reflected in the environmental movement, the State Department acquired more regulatory responsibilities (for example, nuclear nonproliferation, acid rain, global climate change, international fisheries). In the 1980s promotion and regulatory goals were both in evidence. The concern with U.S. competitiveness has emerged more recently as a prominent goal.

The question whether the State Department has understood technical issues well enough to serve U.S. interests remains a hotly debated topic among insiders and the attentive public that follows such matters. The issue is especially thorny since the criteria for success in the integration of scientific with foreign policy expertise have proved so elusive. Calls for closer relations between the State Department and the scientific community have not clarified what precisely the scientist can offer the foreign policy professional.

The State Department neither manages nor funds research programs, nor is it a conventional line or operating department. This has contributed to the relative lack of interest in or familiarity with scientific issues. In practical terms, for many observers the issue of science in the State Department boils down to the status of the OES bureau. Is the OES strong or weak? But this formulation begs the question whether a strong science bureau is an end or merely a means to some larger end (for example, an improved policymaking process or a better foreign policy). In general the OES has performed a number of tasks creditably, carrying out U.S. policy objectives and supporting other bureaus and government departments. Several of the Foreign Service's most distinguished career officers (including Thomas Pickering and John Negroponte) have served as assistant secretaries of OES. The bureau's status and existence seem ensured, even if it is not a front-rank bureau and its influence has been modest in the high policy echelons.

Although the bureau has become institutionalized within the department, there has evidently been some erosion of its technical capacities. The bureau has learned how to function more smoothly with other bureaus at the cost of becoming more like them and being less distinctively technical in outlook. The proportion of scientists and engineers within the bureau has diminished over the past decade. Fewer than one-third of the scientific attachés serving overseas have technical credentials (compared with approximately two-thirds in the late 1970s). At the deputy-assistant-secretary level, very few officials with scientific or technical credentials have been appointed during the past decade. But this may or may not be significant depending on whether one thinks the bureau should be technically oriented or emphasize the more traditional Foreign Service perspectives.

That the bureau has become more "Foreign Service-ized" is a direct result of decisons made a decade ago. As noted, in the mid-1970s Assistant Secretary Dixie Lee Ray, herself a scientist, added technical experts to the traditional Foreign Service professionals. Subsequent disillusionment over the lack of bureaucratic skills displayed by many of those technical people led to a reaction. Senior officials concluded that the bureau's future lay in becoming more closely integrated into the department's normal career patterns and practices. This integration occurred but was incomplete. Science affairs became a "cone" or perhaps a "half cone" in the normal Foreign Service career progression (along with the economic, political, counselor, and administrative cones). The difficulty was that serving a two-year or three-year stint in the OES along with one or more two- or three-year terms as a scientific attaché overseas did not add up to a full career.

Nor has OES service been seen as a prestige assignment despite some improved recruitment. The main problem has remained that the incentives for service at the lower and middle career levels have not been as great as those in the bureaus organized around the traditional area specialties.[11] It has been difficult, in brief, to recruit enough of the most promising junior officers. Until and unless this basic set of incentives changes, the bureau will continue to struggle with its identity crisis and its uneasy compromise between the worlds of science and diplomacy.

8 ||| Science Advisers at the Presidential Level

SCIENCE ADVICE at the White House has been a subject of intense interest to leaders of the scientific community and to the attentive public that follows science policy issues. For many scientists the need for high-level scientific advice has come to symbolize science's importance to the nation. They make the argument that science can render an important service to the nation through a distinguished committee of scientific advisers serving the president; and such an advisory committee presupposes an assistant to the president for science and technology who should act as the chairman of the advisory committee.[1]

The assistant (or science adviser) is the central element. To be effective the science adviser should be assisted by a small staff that advises him and works with the advisory committee and its various specialized panels.

Some observers have viewed direct access to the president as a matter of particular importance. Vannevar Bush, for example, considered that direct access was one of the central factors that enabled the Office of Scientific Research and Development to succeed during World War II. Since then the concept of access to the president has had a remarkable hold on the imaginations of many scientist-statesmen. How often the science adviser sees the president is sometimes seen as a barometer of influence.

This preoccupation with science advice at the presidential level and with direct access to the president is understandable. Specialists in other issue areas (including energy, environment, telecommunications, information policy, drug enforcement, and space) have sought the prestige and presumed importance of White House status. The official guardians of the president's time in his immediate office and in the Office of Management and Budget for their part have fought against easy access to the president. They have periodically sought to

155

prune the number of special assistants and advisory councils in the Executive Office of the President. Institutionally their preference is for a small, high-quality internal staff composes largely of generalists.

The case for high-level science advice requires critical scrutiny. The questions include the following: Does the president need a full-time science adviser in the White House? If so, does the president or the science adviser need a standing advisory committee, part-time or full-time? If a committee is needed, what kind of subcommittee and what kind of staff support are required? And how should the science adviser, the adviser's staff, and the advisory committee fit in with other advisers and staff units in the Executive Office of the President?

This chapter explores these questions and attempts to provide some answers. My answer to the first question is yes, the president does need a science adviser. The presidency has grown too complex for the president to operate with only a handful of generalist advisers, as called for by the 1937 President's Committee on Administrative Management. But of course the adviser's job is not to funnel technical information to the president, who is inevitably besieged with information at all times. The science adviser's job is to express his judgment on complex issues and to help act as the president's gatekeeper and value arbiter. The science adviser, in particular, can help to assess complicated special pleas from interested parties who put their own parochial spin on issues.

And yes, for the same reasons the science adviser needs a staff of modest size to keep track of the vast sea of proposals and ideas flowing toward the center of government.

It is less evident that a standing committee of science advisers will always serve the president's needs. The president should have the kind of advisory body—formal or informal, full- or part-time, small or large—he or she wants. Nor is there a universal formula for the science adviser's job. Unique factors mark every presidential administration. The science adviser's role necessarily must conform to the presidential leadership style. Like most presidential aides, the science adviser will probably provide some mix of policy advice, budget advice, troubleshooting, coordination, and public advocacy. The job will most certainly be more political than technical. Every matter that can be easily resolved by experts will have been resolved by the time an issue reaches the president. There is an analogy to law: matters of fact, law, and values are sorted out by the trial, appellate, and finally the Supreme Court, respectively; the president, like the Supreme Court, deals with the dimension of values. The presidential science adviser's task is thus to search out the hidden value premises and conflicts in the layers of argument advanced by government departments or interest groups.

Does the President Need Science Advice?

Presidents and their immediate advisers have rarely felt quite the same need for science advisers as most heads of agencies, especially the technical agencies.[2] One reason for this has just been stated: presidents deal with issues of value and value conflicts. The science they need to know can be explained by an intelligent and informed lay adviser in the context of the broad policy issue under review. Another reason is that the policy game is different at the presidential level. Senior agency officials view advisory committees as helping to persuade the next decisionmaking echelon of the soundness of their plans. The agency head sees the advisers as a means of building support and convincing the president. The president has no need for this function. The buck stops with him.

Science advice at the agency level is also more important than at the presidential level because the agencies are the initiators and main implementers of policy; the president functions in an appellate mode. Policies must be thought out, programs administered properly, and mature and well-conceived ideas forwarded from agencies to the president.

Virtually everything that touches the White House is political in one of several dimensions: the president as the party leader who must watch out for appointments, elections, and issues that influence the fate and vitality of the national and local party organizations; the president as legislative leader and primary negotiator with Congress on a range of interrelated policy issues; the president as ceremonial leader and communicator with the general public and its many specialized subcommunities; and finally the president as the one who articulates the aspirations of the American people and sets the overall tone and direction of the high politics of the era.

Moreover because the president has broad political responsibilities, the White House has extra sensitivities connected with advice and advisers. If the president is to have a science adviser, he or she must be absolutely committed to the president's agenda and understand that anything he or she does will reflect on the administration. By the same token the creation of any advisory committee will involve extremely sensitive political considerations. The president may find an advisory commission useful to educate the public or Congress, or both, but there are many other avenues potentially available for this purpose that may be more effective and potentially less risky.

The advice side in the multidimensional role of the advisory committee is less important for the president than for the executive agency head (even here, providing support may outweigh providing advice as the motive for creating a committee). The president as the focal point of the executive branch is unlikely to overlook entirely a salient issue or miss critical technical dimensions of a problem. By the time any issue comes to the president, a great deal of staff time in the agency level, the OMB, interest groups, and other outside sources has been spent on the problem. The president and his closest advisers will have been inundated with advice of all kinds. The problem is that nearly all the analysis of the issue that has been done previously has been mixed with institutional loyalties, self-interest, and parochial outlooks. The president needs broad judgment that is removed from agency parochialism or special interest pleading.

The kind of advice the president most needs concerns the complex interplay of issue substance with politics and the relation of any set of policies to the larger goals of the administration. Those best equipped to advise the president on such issues are political generalists who can integrate enough knowledge of policy substance with the intricacies of political timing, nuance, and aspiration. Such individuals may be scientists, lawyers, engineers, or political professionals by background (scientists are often very good politicians). But the function performed is different from virtually anyone's professional specialty or training. As has been said, there is no cram course for the presidency or for being a close presidential adviser.

Many agency-level technical advisory functions—peer review of grants, advice on management of large technical programs, regulatory functions requiring a technical underpinning—simply do not exist at the White House level. The White House is thus a poor candidate for the kind of advisory role that scientists find comfortable, peer review of a scientific or technical activity requiring expertise and research background in a particular field.

The science adviser at the White House level must be extremely broad-gauged, have knowledge of the R&D priorities of numerous agencies, and understand politics and the fishbowl atmosphere of

Washington. The issue areas where the science adviser can help the president include, among others, assessing the overall performance of technical departments, sorting out competing claims from departments on issues with technical content, and anticipating emerging problems. In short, the president needs a science adviser who is politically compatible and experienced and who adds depth and wisdom to a range of issues where the president feels uncomfortable with what the agencies or program advocates are telling him. The adviser must also be a trusted source of confidential counsel as well as an articulate public spokesman for the president's programs. Advisers who are generalists will of course remain crucial to any president. But presidents need substantive staff aides as well.

If there is a strong case for a full-time science adviser (and for a staff that enables the adviser to function), does a compelling case also exist for a committee of presidential science advisers? If so, what kind of advisory tasks or functions would such a committee perform? An initial answer is that useful advice will relate to how to deal with new and pressing problems not being addressed by the departments or how to improve or manage existing programs that are in serious trouble. The ad hoc or temporary presidential commission may be a more natural vehicle to the president and close aides than a standing advisory committee. But the standing committee offers some advantages as well: a special panel can be speedily assembled under the umbrella of the larger committee to focus on new developments; the president will not necessarily want to create a special commission for each matter that engages his attention; the practical problems and formalities of creating a committee under the Federal Advisory Committee Act (FACA) are a disincentive to new committee creation; and not every issue is of sufficient importance to merit the creation of a full-blown commission.

The science advisory committee provides a useful way to identify and marshal needed expertise quickly. A good example of such a role is the 1960s installation of a new air traffic control system for the nation. The President's Science Advisory Committee (PSAC) panel, which included non-PSAC members and which worked closely with a Federal Aviation Administration advisory panel, recommended the scrapping of a three-dimensional radar system in favor of requiring a transponder in each aircraft. Another 1960s panel proposed the system of high-altitude placement of communications satellites that now exists, as opposed to the low-altitude system AT&T preferred. The

PSAC panel also had an impact on the Communications Satellite Act that Congress adopted in 1962.

There may be occasions, on the other hand, in which a special commission seems called for. The report of the Augustine committee on the future of the U.S. space effort is a good illustration of the ad hoc presidential commissions. The National Aeronautic and Space Administration's problems had reached such a crisis stage that a special full-dress treatment, with attendant publicity and visibility, was deemed necessary (see chapter 6).

The case for a standing advisory committee to serve the president rests principally on the ability to provide expertise as needed, quickly and with flexibility. Presidents will wish to supplement such a standing committee with temporary commissions when a special, high-profile focus seems desirable. Presidents will find many alternative means to obtain advice and to elicit support; they will not and should not feel bound to rely on any particular advisory structure. The political sensitivity of any White House action means that politics will always be a part of the creation, composition, and functions of any committee operating at the White House level. So far I have discussed the matter in terms of administrative logic and theory. But the reality of the president's situation can perhaps be best understood by posing the following questions: What has been the experience to date with science advisers at the presidential level? How have presidents and their closest advisers viewed the role of the science adviser? How useful have they found the adviser and the formal science advisory committees?

A Brief History of Presidential Science Advice

Several general points stand out from a look at the history of presidential science advising. First, scientists have played useful roles during wartime. Certainly during World War II and its aftermath scientists contributed to critical decisions. This was not true during the Vietnam War mainly because scientists opposed the war, or at least Presidents Johnson and Nixon believed they did. This belief led to the exclusion of the science advisers from national security decisionmaking, and for a time to the abolition of the post of presidential science adviser. Second, science advisers have not succeeded in being as useful in the domestic policy arena. As Edward Burger concludes, "The liaison between science (and scientists) and the public policy machinery at the corporate level of the government is a difficult and tenuous one. . . . The mandate of science for policy making has simply not been fulfilled."[3]

FDR and the New Deal

President Franklin D. Roosevelt, as the first chief executive of the modern era, is an appropriate beginning point. His term in office, which saw the creation of the Executive Office of the President and the emergence of a large-scale federal bureaucracy, also witnessed the first effort to create a science advisory committee in 1935. This preliminary effort led to clear failure. The embryonic effort, among other difficulties, ran into the problem of squabbles within the scientific community over which fields should be represented. It ultimately failed when "soft" scientists, that is, economists and social scientists functioning as close presidential advisers, concluded that such a committee had little useful advice to offer the president (or what amounted to the same thing in practical terms, they did not wish to have a new group interfering with their own role). One key problem was that members of the Science Advisory Board were indifferent to the political environment in which they operated. Political generalists could not readily see how science could provide workable answers to the country's problems. Lewis E. Auerbach sums up the effort in these terms: "Consideration of political *realities* was alien to most Science Advisory Board members, who wanted politicians to appreciate the needs of science but who had little interest in the desires or needs of politicians."[4]

There was a general strain in New Deal thinking that viewed science with suspicion. Science was seen as causing the Depression because overproduction had resulted from the application of labor-saving production techniques.[5]

As war clouds gathered, the position of the scientists shifted. FDR benefited from scientific advice first in the form of the famous Einstein-Szilard letter warning of a potential Nazi atomic bomb.[6] The case could be made that President Roosevelt might have made more expeditious use of that advice had a science adviser or a science advisory committee existed within his official family. But absent the international crisis neither FDR nor his closest advisers had felt the need for such an office. The President's Committee on Administrative

Management in 1937 (the Brownlow committee), which laid the foundations for the Executive Office of the President, recommended that six aides serve the president, all of them generalists and all having "a passion for anonymity."[7]

The fathers of public administration theory did not regard science as being of enough general importance to merit a place in the family of presidential advisers. Moreover, the passion for anonymity could contradict a key tenet of the scientists, namely, the importance of a committee of outsiders working only part-time and having their main professional base in a university or other nongovernmental setting.

The war brought dramatic changes in this line of thinking. Vannevar Bush functioned as a presidential science adviser during the war. He was, however, more closely involved with the administration of the defense technical programs and with the civilian and military leaders in the Departments of War and Navy than with providing broad policy advice to the president. His famous report, *Science—The Endless Frontier*, in which he laid out his vision for postwar science organization, came not at his initiative (though he did formulate the questions and draft the presidential letter initiating the review). The idea for the report came from a political adviser to FDR, Alexander Cox. Cox's conception was to have a report to be used as ammunition against the Republican nominee in the 1944 presidential election. Vannevar Bush refused to countenance this political use and persuaded Cox that the report should be deferred until after the election. It was submitted to President Truman in July 1945, after FDR's death in April.[8]

The Early Postwar Years

As seen in chapter 2, science advisers began their service after the war at the agency level following extensive involvement with the military services during the war. The first science advisory committees were in the military services and the Atomic Energy Commission. They initially concerned the content of R&D programs. In 1950 William T. Golden, a New York investment broker and aide to AEC chairman Lewis Strauss, was charged by President Truman to study the overall effectiveness of the military's technical programs. Golden became concerned that there was a need for an effective science advisory apparatus at the White House level. In his report to the president he recommended the creation of both a full-time science adviser post and a part-time advisory committee.[9]

President Truman accepted the recommendations in part. He created a science adviser and an advisory committee within the Office of Defense Mobilization (ODM) under the direction of General Lucius D. Clay. He stipulated that the committee would also "from time to time" advise him. He did of course have access to other sources of advice on technical issues, including the AEC's General Advisory Committee (whose advice he rejected when he decided to approve the development of the fusion bomb in 1949) and the director of the Bureau of the Budget (whose advice to veto the National Science Foundation bill in 1947 he accepted). President Truman did not in practice make much use of the ODM science advisory committee.

The ODM committee chafed under the supervision of General Clay. After Oliver Buckley, the first chairman and initial ODM science adviser, resigned in June 1952 because of ill health, the committee sought to have itself upgraded so as to report either to the president directly or, failing that, to the director of the Bureau of the Budget. The effort was rebuffed. Discouraged, committee members considered resigning en masse.[10] But they finally decided to wait out the election of 1952.

The Eisenhower Era

Science advisers began to play a useful role in the first term of the Eisenhower presidency when the ODM science advisory committee, chaired by I. I. Rabi in the mid-1950s, started to function effectively in White House–level science advice. That the ODM committee still lacked formal designation as a White House unit was irrelevant. The committee's mission was decided by Eisenhower, and its business was conducted on his terms. Rabi and his colleagues did not worry about reporting channels or about where they sat in the organizational hierarchy. They perceived themselves as public servants in the old-fashioned sense: they displayed the "passion for anonymity" that the President's Committee on Administrative Management had called for in presidential advisers; the committee did not embarrass or frustrate the president or his closest aides; no one postured before the press or Congress and no one attempted independent political maneuvers.

The committee convened temporary panels to conduct several important studies. The Killian panel on technical capabilities underlying the nation's strategic posture and the Gaither committee on missile defense were two prominent study efforts (the former was highly influential but the latter, opposed by Eisenhower, was quietly buried). The von Neumann committee, an early effort dealing with antiballistic missile defense, was also organized with the assistance of the ODM committee. The committee avoided conflicts of interest by focusing solely on policy issues, leaving to others the task of articulating science's own needs.

Various factors account for the success of the ODM committee during this period. The most important was simply that President Eisenhower wanted the advice. Other important contributing factors were the right personal chemistry, the emergence of pressing issues with a heavy technical content, inadequate technical competence within the regular departments at this time, an inadequately developed staff system to deal with such complex issues in the White House itself, and the willingness of advisers to put in extensive effort without self-aggrandizement or the search for public acclaim.

That there were no later conflict-of-interest and open meeting requirements to constrain the president's access to outsiders also helped to make possible this period of freewheeling advisory activity. This early effort by the ODM committee contributed to the nation's early planning for strategic air defense, missile development, and strategic force posture. Virtually none of the pre-1957 advice to Eisenhower was directed toward the interests of the scientific community itself, that is, toward the development of new research fields, support for scientific facilities, and the like. The effort closely approximated President Tyler's criterion cited in chapter 2: the president's desire to inform himself and to do so confidentially, reserving the right to decide when, how, and if wider dissemination of the advice should be made.

The development of a technically sophisticated National Security Council staff had not yet occurred. Eisenhower evidently felt more comfortable with his science advisers than with General Robert Cutler, his national security adviser, on the technical aspects of national security and space policy.

The high point of science advice in the White House is conventionally dated from the formal creation of the science adviser position and the PSAC in October 1957 after the Soviet launch of *Sputnik I*. President Eisenhower appointed James R. Killian, Jr., then president of the Massachusetts Institute of Technology and an able and active participant in the ODM advisory effort, as the first presidential science adviser. The period from the creation of the science adviser post in the Eisenhower administration until President Kennedy's assassination November 1963 is the period that most observers have deemed the golden age of presidential science advising.

From the outset Killian directed his attention largely to matters of national security, especially to strategic forces planning, air defense, atomic energy, and the military uses of space. As time passed, the interests of the scientific community itself—the need for research funds, student support, and facilities—were gradually added to the science adviser's agenda. The PSAC members and scientists in general, as well as politicians and other national leaders, had come to believe that expanding the country's educational and research resources was a vital national goal. This shift in emphasis added a new element: science advisers would deal with issues that could be viewed as advocacy for science (for example, agency R&D budgets, funding for basic research, government relations with universities, and the like).[11]

Once the nation's strategic posture had been laid down in the triad of land-, air-, and sea-based systems by the end of the Eisenhower presidency, it became less obvious that the scientists possessed unique policymaking expertise or filled a special need. Thereafter refinements in force posture, the integration of new technologies into the operational inventory, and the political uses of military force focused less strictly on the technical characteristics of systems. Splits within the scientific community between the "peace-through-strength" and the "peace-through-arms-control" advocates also diluted somewhat the luster of the scientists' wartime service and the popular faith in science. It was perhaps inevitable that scientists gradually would become less unique as sources of national security expertise. As others became familiar with the strategic issues, the scientists had to share influence. Moreover, as the scientists confronted a new range of issues in domestic and social policy that seemed less amenable to the hard sciences, their relations with other members of the White House staff were bound to change.

Another factor was the relative influence of the full-time staff in the science adviser's office versus that of the part-time advisory committee. Some full-time staff began to view the part-time body as burdensome; it lacked up-to-date information on fast-breaking issues and had to be briefed constantly. The full impact of the various erosive trends was to be felt only later, however. During the Eisenhower presidency the science office remained influential. Killian (1957–59) and George B. Kistiakowsky (1959–60) enjoyed easy access to the president and provided him with useful advice on a range of defense, space, atomic energy, and other policy issues.

The science advisory system in the Eisenhower period can be credited with playing an important part in the nation's basic strategic doctrine and in the management approach to ICBM development, the organization and development of national space policy, the establishment of the Department of Defense Research and Engineering office and the Defense Department's Advanced Research Project Agency, and the initial thinking that subsequently led to the creation of the Arms Control and Disarmament Agency (ACDA) in the Kennedy administration.

The Kennedy-Johnson Years

A new period began with the election of President John F. Kennedy and the appointment of Jerome B. Wiesner of MIT as science adviser. The special relationship that Wiesner enjoyed with Kennedy was hailed as an indication of a continued influential role for scientists in the new administration. In actuality, Wiesner's term as science adviser showed the vulnerability of the scientists' position and was marked by a gradual erosion of influence. The science office proved to be fragile and exposed in the struggles for status, the shifting chemistry of interpersonal relations, and the political crossfires of the Kennedy administration.

Wiesner's troubles began as early as April 1961 when he angered the president by purported unauthorized disclosures of information to Russian colleagues at the time of the Kennedy-Khruschev summit in Vienna. Wiesner was gradually frozen out of national security issues, the very area that was the centerpiece of the scientists' effort in the Eisenhower administration. Indeed, Wiesner had virtually no role during the Cuban missile crisis of October 1962. The emerging dominance of the national security adviser and NSC staff gradually reversed the relationship that had prevailed under Eisenhower; the science office became the junior and the security adviser the senior partner. Wiesner and security adviser McGeorge Bundy worked together cooperatively, however, and shared staff members on disarmament and arms control issues. Wiesner played a role in the Partial Test Ban Treaty of 1962 and in the creation of the ACDA. His increasing focus on arms control put him in conflict with the Defense Department and with Kennedy aides who favored a more cautious approach to disarmament. The scientific community was increasingly seen as divided between the "peace" scientists in the academic community who dominated the scientific establishment and the "war" scientists from the national laboratories.

Conventional wisdom has tended to view the act of persuading President Kennedy to issue Reorganization Plan No. 2 in 1962 and thus institutionalize the Office of Science and Technology (OST) in the Executive Office of the President as an important achievement. Science henceforth would have a "permanent" status in the White House. The move was actually a sign of the OST's fragility. Wiesner was afraid that without a formal base the advisory system might disappear in the crosscurrents of politics. The OST did lack the stable resources needed to function effectively had it remained under the president's discretionary funds. Formalizing the OST solved this problem but added to forces pushing the science advisory role toward the periphery. The science adviser was now concerned with the distractions of managing an office. He could be called by congressional committees to testify in his capacity as OST director, a practice at odds with the norm for the president's closest advisers.

Reconciling the multiple roles of the science adviser—manager of the OST, chairman of the PSAC, intimate presidential adviser, adviser to the Budget Bureau on science budgets, visible spokesman for the scientific community—became more difficult in the Kennedy years. What suffered most was the role of intimate presidential adviser. It became increasingly apparent that the science adviser was not in the inner circle of presidential advisers.

When Wiesner expressed his desire to resign in 1963, he recommended chemist Donald Hornig from Brown University as his successor. The assassination of President Kennedy occurred before Hornig formally took up his duties. Hornig was not personally close to President Lyndon B. Johnson. For a time it was uncertain whether Johnson would retain Kennedy's choice. Johnson decided to stay with Hornig because of the importance of maintaining continuity in government and of reassuring the nation in the wake of the assassination.

As the controversies surrounding the Vietnam War intensified, LBJ's relations with his science advisers, never close, grew more distant.

Publicly expressed opposition to the war by scientists who were serving as government advisers became a particular source of irritation to
Johnson.[12] Johnson expected loyalty from his advisers, all of them.
The last impression he wanted to give the public or Congress was that
the executive branch was a house divided. By the same token, he had
little appetite for an open break with the scientific community. So
rather than dissolving the PSAC and banishing the war critics from
his official family as he would have like to do, he ignored them.

The displeasure with the PSAC and with the scientific community
generally carried over to LBJ's relations with Hornig. The president's
science adviser found himself increasingly isolated and caught in the
middle between the scientific community and the president. It was an
uncomfortable role that Hornig bore with grace and dignity. A small
but telling sign of LBJ's attitude toward the PSAC was his decision in
1968 to ban committee members from eating in the White House mess.
By the end of the Johnson administration the science adviser no longer
played a consequential role in national security policy.[13]

That Johnson's science advisory committee was at odds with his
administration seems in retrospect unsurprising; at that time, the entire scientific community was rent with deep and sometimes bitter
divisions. Although the disagreements did not follow any simple fault
line, the war played a large part in many of them. In the past it had
been assumed that science could serve equally the needs of national
defense, economic growth, improved standards of living, and better
health for Americans. Vietnam was eroding this assumption and frac-
turing the postwar consensus that defense expenditures were compatible with and even contributed to a flourishing civilian economy.[14]

All of this dovetailed with other forms of unrest that appeared in
the late 1960s and early 1970s. Mounting skepticism among consumers about product safety and reliability, the gap between the established culture and the so-called counterculture, and concern about the
impact of industrial activity on the environment, all affected how the
public perceived science as well as how scientists participated in policymaking. Science and technology came to be viewed as the source of
many ills, including toxic waste, air pollution, and occupationally in-
duced cancers. The fuel for these suspicions came for the most part
not from politicians but from faculty and disaffected students and from
whistle-blowers, critics within the scientific community itself.

The Fall and Rise of the Presidential Science Advisory System

The PSAC, which represented a cross section of the scientific community, was as plagued with controversy as was the broader scientific community. From Johnson's term of office onward PSAC members had difficulty keeping their disagreements, both among themselves and with the administration, from spilling over into the public arena. The rules of the game for science advisers, never precise, became increasingly confused. Some science advisers felt they were being used less as a source of recommendations than as a way of lending support to the government's policies. Believing that their silence would be misconstrued as approval of those policies, science advisers spoke out against practices and programs with which they disagreed.

Tensions between the president and his science advisory committee continued until President Richard Nixon, angered that some PSAC members and consultants had publicly opposed two of his favorite projects—the construction of a supersonic transport and the development of an antiballistic missile defense system—dismantled the science advisory apparatus (more precisely, he removed most of its functions from his immediate office) in January 1973.

The science advisory role during the Nixon period has been one of the most misunderstood chapters in the postwar history of government-science relations, heavily overlaid with partisan commentary. It is instructive to get the story straight.

At the start of the first Nixon term, there were high hopes for an influential role for the science adviser and for scientists. Lee A. DuBridge, a respected and experienced scientific statesman and president of the California Institute of Technology, knew President Nixon personally. As science adviser he would presumably enjoy easy access to the president. But several early decisions limited DuBridge's role and for practical purposes doomed the office almost from the start. One decision was beyond the science adviser's control: Melvin Laird, the former Wisconsin congressman tapped for the secretary of defense position, made it a condition of accepting the job that the science adviser be excluded from national security matters.[15] Once he assumed office Laird made sure that DuBridge was kept in a marginal position. The science adviser was now virtually excluded from the area where in the 1950s the science advisory system had been most influential.

The second decision was DuBridge's own. At the start of the administration he injected himself into the budget process on behalf of the National Science Foundation in a way that angered the budget director, Robert Mayo, and his close associates. DuBridge finally won a $10 million increase in the NSF's appropriation, but it proved to be a Pyrrhic victory. The budget officials saw the intervention as special pleading for science and thereafter froze DuBridge out of the budget process. With the science adviser in a marginal position, the rest of the advisory apparatus became virtually functionless. Only through an effective science adviser can the part-time advisory committee play any meaningful role.

There were other areas in which the science adviser and his office sought to play a role: health, environment, energy, and others. The OST and the PSAC achieved a modicum of influence in these areas, mostly by lending the prestige of science and the scientific community to policy initiatives emerging from the federal bureaucracy.[16]

When, however, a solid base of scientific expertise and advice was lacking at the agency level—as was the case, for example, with the early program efforts of the Environmental Protection Agency—the staff units serving the president could not remedy the deficiency. The OST's efforts to inject "good science" into the policy debate over regulation of 2,4,5,-T in the EPA's early period, for example, had no effect.[17] The presidential-level advisory capacities could not, in short, make up for the inadequate staff work at the agency level.

The OST could not even effectively influence many of the early environmental decisions made at the White House itself. The creation of the EPA as an independent agency and the clean air amendments of 1970, for example, were essentially political decisions made on the basis of perceived public impact. The president wanted to show strong symbolic commitment to environmental goals and to recapture the initiative from the Democratic leadership in Congress.

DuBridge's early resignation was unsurprising under the circumstances. His successor, Edward E. David, Jr., a forty-five-year-old computer specialist from the Bell Laboratories, had no illusions about being a member of Nixon's inner circle or having major influence in policy. He sought to improve relations with George Shultz, the director of OMB. He tried (but failed) to persuade Nixon to pursue a less expensive version of the space shuttle, advised against the joint Apollo-Soyez mission on the grounds of risk (but was overruled by Kissinger), supported the administration's war on cancer, won some modest increases in the NSF's funding, and generally showed a pragmatic ability to get along with other White House staff.

Meanwhile the Vietnam War protests continued to haunt the nation as the Nixon administration struggled to negotiate a settlement. The conflicts between the scientific community and the administration that had bedeviled President Johnson intensified under Nixon. They were exacerbated by the presence on the PSAC of holdover appointees and consultants-at-large from previous Democratic administrations.

There was a theory at the time that continuity was important for certain government scientific positions. The director of the NSF was one such position. When President Nixon initially rejected DuBridge's choice of Cornell chemist Franklin Long to be the NSF director, he did so on the grounds of Long's opposition to the Antiballistic Missile (ABM) program.[18] The presidential science advisory committee was also believed to embody neutral expertise, at least according to the informal rules of the game as understood by the scientific community. To inject partisan considerations into the appointment process for the PSAC, for example, was thought by the scientists to be inappropriate.

The politicians had a different interpretation of the informal rules: everything at the White House reflects politics and the values and aspirations that the president seeks to promote. Democrats in the odd position of consultant-at-large to the PSAC could use their privileged position within the White House family to gather information and snipe at administration policy. This seemed an intolerable anomaly to Nixon and his political aides. The remedy was to make sure the science advisory apparatus had no genuine role in policymaking. Henry Kissinger, as national security adviser, found another tactic. He appointed and used his own science advisory committee while excluding DuBridge and PSAC members from deliberations on the ABM issue, the Multiple Independently Targetable Reentry Vehicle (MIRV) controversy, or the technical verification issues concerning the Strategic Arms Limitation Talks. When Kissinger eventually became convinced that even his own panel was not supporting administration policies, he stopped meeting with them.

The Nixon science advisory committee was therefore virtually without influence from the start, though it continued its formal existence. An experienced science adviser and member of the Nixon committee summed the matter up in these terms: "The PSAC was still held in high esteem by the scientific/technical community. Unfortunately, neither the power structure within the White House nor, indeed, the various departments and agencies of the government shared this attitude. As a result, the Committee was largely ineffective."[19] The reasons for the PSAC's ineffectiveness were quite simple: "Most— if not all—of the work of the Nixon PSAC was self-initiated, so much so that we were meddling in the internal affairs of government departments; our work was not of burning interest to the policy- and decisionmakers who then served in the government."[20]

The basic reasons for the demise of the science advisory apparatus can be stated simply: the political outlook of the scientists was just too far out of sympathy with the Nixon administration for the arrangement to continue. Just as science advice did not take root under Roosevelt because the scientific leaders were too Republican and conservative for the president and his advisers, the scientists now were too Democratic and liberal to be part of the inner workings of the administration.[21]

The two episodes that illustrate the general point and that probably more than any others caused the final demise of the advisory system were the disputes over the ABM program and the supersonic transport (SST). The ABM issue was a problem for the Nixon administration from the start. The PSAC and the science office were pointedly excluded from the debate over the proposed *Sentinel* missile defense system in large part because the anti-ABM views of PSAC members were well known. A Defense Department review in late February 1969 resulted in a report to President Nixon and security adviser Kissinger recommending that a new ABM system, dubbed *Safeguard*, be pursued that would focus on the defense of the nation's intercontinental missile sites instead of thin defense of cities.

In early March 1969, hearings before the Senate Foreign Relations Committee chaired by Senator J. William Fulbright (Democrat of Arkansas) launched a critical fusillade against the ABM by a succession of PSAC alumni and former science advisers. James Killian urged the creation of a committee like the Technical Capabilities Panel he chaired under President Eisenhower to review the ABM. Kistiakowsky, Hornig, Wiesner, and others criticized the technical flaws of the ABM and its destabilizing effects on the arms race. On March 4, 1969, in the middle of the hearings, Nixon announced that he would seek additional funding for the new concept of defense of missile sites embodied in *Safeguard*.

The opposition of the scientists, aligned with administration critics in Congress, was finally overcome, but only after strenuous efforts by the administration. The Senate approved *Safeguard*'s initial appropriation by only one vote in the summer of 1969 and finally in mid-August 1970 defeated a last-ditch effort to restrict funding for the program by 52 to 47.[22]

The ABM controversy had serious implications for DuBridge's role in the Nixon White House. Like Hornig on Vietnam, DuBridge on the ABM found himself caught between his colleagues on the PSAC and the president and the national security adviser. In an effort to show scientific support for the administration's plan, DuBridge on March 17 made public a letter of support to the president; he had attempted to obtain the signatures of PSAC colleagues but failed to do so. Nixon and Kissinger were angered at the lack of support from the scientists, and Kissinger was reinforced in his belief that the committee was filled with administration critics operating under the pretense of scientific neutrality. The president was also angered by a leaked story in the *Washington Post* on the same day he announced his ABM decision. The story depicted the administration as ignoring scientific advice from every source except the Pentagon in the ABM decision.[23]

The protracted battle over the ABM had the effect of eliminating the PSAC from any role in other critical defense decisions such as the MIRV issue, which turned in part on the technical assessment of whether satellites could detect Soviet missile launches. The scientists, who saw themselves as the essence of objectivity, were seen by the administration as volatile, emotional, and highly biased. Indeed they were viewed as stalking horses for Democratic critics in Congress. To the scientists the ABM episode was proof that the administration did not want advice but merely public support. To Kissinger there was the problem that the whole process of science advising "was becoming chaotic. For the president to make a decision we had to offer him some general concepts related to our national security, rather than make him arbitrate exceedingly technical controversies."[24]

The SST episode was the final straw. The controversy had its origins back in September 1963, when President Kennedy outlined a goal of building "a commercially successful supersonic transport superior to that being built in any other country." This goal was to be achieved by the end of the decade. President Nixon, upon taking office, asked the Department of Transportation to create an interagency SST review committee, its members to include DuBridge as well as representatives from other departments. DuBridge shortly created an ad hoc SST review committee of his own with the president's blessing and made PSAC member Richard Garwin, an IBM physicist, its chairman.

Both committees issued critical reports. The Garwin panel—citing numerous unresolved questions relating to noise, commercial viability, and the environment—recommended outright cancellation of the project. The Transportation Department's interagency study was less harsh, and a summary prepared by Under Secretary of Transportation James Beggs for Nixon contained some mild praise for the airplane.

In September 1969 Nixon announced his decision to go ahead with the SST, ignoring the recommendations of the Garwin report. He released to the public only the summary of the Transportation Department's study. In the wake of the Nixon decision, Garwin sent a letter to each member of his panel stating his views on dissent: "I believe that the communication in any administration, and in particular in the present one, is sufficiently poor, and the need for education of responsible officials so great, that it is unwise for high-level advisers to restrict their advice to one person."[25]

Garwin later was called on to testify before Congress on the SST when the existence of the panel report was inadvertently revealed in testimony by an administration witness. Garwin agreed to testify on "technical" issues relating to the SST; in appearances before congressional committees in April, May, and August he, in effect, presented the full argument of the panel.[26] Congress nonetheless in 1970 voted funds for the SST project for another year. In the following year, after mounting opposition, Congress killed the SST in one of the first instances of a major technological project being blocked for environmental reasons.

Garwin's decision to testify against the administration's position on the SST while still a member of the PSAC touched off a bitter controversy. Members of the committee, former science advisers, and the wider scientific community were divided on the wisdom of Garwin's action. DuBridge later cited Garwin's action as part of the reason for the PSAC's demise (DuBridge himself had opposed the SST in private but had backed the administration's policy once the president had made his decision). Others believed that Garwin should have resigned before his testimony. Garwin had defenders on the PSAC and in the scientific community; a poll of scientists showed about half thought Garwin should have resigned first before going public with his criticisms, and half defended his actions.

If scientists were divided in their assessment of the informal rules of the game on dissent, the White House reaction was clear—the PSAC and the scientists had grossly transgressed the tacit understanding that should govern advisers serving the White House. The committee had in consequence become a political embarrassment. One White House staffer declared, "Who in the hell do these science bastards think they are?" Another said, "Who needs this bunch of vipers in our nest?" Yet a third member of the White House staff publicly declared that he now regarded the post of presidential science adviser as "an anomaly."[27]

Nixon resolved to rid himself of the entire science advisory apparatus. He did so at the conclusion of his first term by accepting the pro forma resignations of all PSAC members and appointing no new ones. Science adviser Edward David resigned his post on January 3, 1973, to return to private industry. It was a bleak moment for those inside and outside of government who believed that the nation needed the advice of scientists at the presidential level.

While politics played a large role in the transfer of the science office outside the White House, there was also an organizational argument. President Nixon had the support and encouragement of OMB director George Shultz in his move. Shultz argued that a separate science office was no longer needed in the Executive Office. The OMB had considerably augmented its own staff resources in science affairs, adding technically trained people who could evaluate agency R&D budget requests and in general assess agency claims on complex technical issues. Science advice, he argued, should be integrated with budgetary and other considerations in policy formation. In his view the OMB was the logical entity to perform this function.

The science adviser post was not actually abolished but was transferred to the National Science Foundation. Reorganization Plan No. 2, which had created the Office of Science and Technology in 1962, had transferred certain NSF functions to the OST. Shultz now took the view that the functions should be transferred back to the NSF. The NSF could advise and assist the OMB and thus indirectly aid the president. The science adviser post in its new incarnation was to avoid involvement in national security policy unless specifically authorized by the president to undertake an assignment. The interagency coordinating function that had been part of the science office's operations was retained at the White House level.

The general reorganization plans stemming from the work of the Ash Council to streamline the Executive Office of the President also played a part in the decision to transfer the science office. Traditional OMB doctrine had always resisted the concept of specialists and special assistants for particular causes in the Executive Office. The OST's efforts to push for expanded R&D programs in the small agencies, and for enhanced research budgets generally, were viewed with suspicion by budget examiners.

The passage of the FACA, though seemingly a minor event, was also an incentive for removing the science advisory committee from the Executive Office of the President. Budget deliberations were supposed to be highly confidential, and open meetings could jeopardize the sanctity of Executive Office processes. OMB management doctrine thus combined with political considerations to help sink the science office. A useful perspective on the whole OST experience is David Beckler's analogy of the heart transplant: "The 'alien' science and technology mechanism functions effectively only as long as the forces of rejection are continually suppressed. In the end, if the environment does not become more favorable to the acceptance of specialized advice, the transplant fails."[28]

NSF director H. Guyford Stever, who now assumed the duties of science adviser along with his normal responsibilities, continued to have access to the president but on a relatively infrequent basis. This, however, differed little from the pattern that had largely prevailed since the mid-1960s. Killian and Kistiakowsky had enjoyed close contact with President Eisenhower, but the relationship became more distant with Presidents Kennedy, Johnson, and Nixon.

Stever's role in opening contacts with Soviet scientists played an important part in the Nixon-era détente. The staff that Stever built up in the NSF to deal with science policy issues numbered about fifty, considerably larger than the approximately twenty professionals staffing the White House OST in the 1960s. Stever's work in energy policy played a part in the national response to the energy crisis of 1973–74. Under the circumstances, with the White House increasingly caught up in and paralyzed by the Watergate scandal, the science advisory unit probably was better off out of the White House. In any event, the nation functioned without the benefit of a White House science adviser until the final months of the Ford presidency. At that time, Stever stepped down as NSF director and assumed the duties of the newly reconstructed presidential science adviser.[29]

In the fight between Nixon and the scientists that led to the elimination of the White House science office from 1973 to 1976 (and of the PSAC for a longer period), it serves little purpose to argue about who was right. By battling against the ABM and the SST, did the scientists prove to be right in the end because the nation got an ABM treaty and did not pursue the SST? Was American democracy well-served by technical experts bringing facts out into the wider public debate that were suppressed or distorted by the administration? Some scientists have felt that even losing the services of the science office was a price that had to be paid for speaking truth to power on issues of high national importance.

Put aside for the moment the question of how one assesses the opportunity costs: the lost chances to influence decisions on other issues, such as on MIRV missiles or SALT verification. Put aside also the question of what issues are important enough to merit breaking the informal rules of the game. Surely the central point of the debate is that there are no unambiguous and objective facts known only to the expert, the suppression of which imperils democracy or the release of which will transform the entire policy dispute. The critical element is the exercise of judgment in assessing the significance of uncertain technical data against a background of value-laden considerations. The White House context will inevitably put a premium on timing, the integration of separate issues into a whole, and the fit between politics and issue substance. Science (and scientists) cannot resolve the kind of public policy questions posed in the ABM and SST cases, because any presidential decision includes judgments on matters that lie outside the realm of science. Practical politicians need not "accept the idea that there are aspects of knowledge which no science can touch. All they need to understand, in order to appreciate the limitations of science as a guide to policy, is that each complex problem involves aspects to which several scientific disciplines are relevant, and that a master of one discipline is not likely to be reliable as a synthesizer of all of them."[30]

The Nixon-era PSAC insisted on aggressively pursuing its own agenda and on second-guessing the operating departments even when it lacked the support of the president. Criticizing departmental policy initiatives is an important potential function of the science adviser but not against the president's wishes and certainly not with the intent to undermine specific administration policies. Departments always have biases in their technical information but so do all advisers. A public fight in which one side's experts declare the other side's experts to be biased or technically deficient—and in which there are rarely objectively verifiable "right" answers—is a disservice to democracy. The inside adviser can never join in the process of public debate for long before losing the status of confidential adviser.

Playing the role of inside adviser has become increasingly difficult at the presidential level because the media have become adept and relentless at ferreting out minor cracks in an administration's policy front and because partisanship has become increasingly intense in American politics since the Vietnam era. The Nixon-era science advisory battles appear in retrospect so dramatic because the president and his close advisers partially believed in the myth of value-free science and only gradually and painfully asserted their authority.

In May 1976, when Congress resurrected the position of science adviser and created the Office of Science and Technology Policy (OSTP), it chose not to recreate the PSAC. The decision to omit the advisory committee from the reconstituted science office was a reflection of the shadow cast by the previous squabbles, coupled with fears that the FACA would make a confidential advisory committee difficult to operate.

From Frank Press on, science advisers made totally clear that they served the president and did not represent the scientific community. Press served President Carter effectively without benefit of a formal advisory committee (but with ample access to informal advice from the scientific community). On the other hand, Ronald Reagan's science advisers, George A. Keyworth II, John McTague, and William R. Graham, were criticized by their colleagues for their alleged reliance on too narrow a base of outside advice, for not being sufficiently influential, and indeed by some for accepting the position in the first place without insisting on direct access to the president. Keyworth served for a time without the benefit of a committee but in his second year created the White House Science Council. The council was smaller than the former PSAC and reported directly to Keyworth rather than to the president.

The Reagan science advisers found a formula for dealing with the potential problems posed by the FACA. They simply held a brief public session and then closed the meeting, citing one of the permitted exclusions (usually national security).

Keyworth gained influence because of the Reagan administration's faith in science and its desire to increase R&D budgets. And in particular his enthusiastic embrace of the strategic defense initiative program earned him credit with his White House colleagues. The impetus for the SDI had not come from the science office; the science adviser was consulted about the initiative only shortly before the president's March 21, 1983, speech announcing the program.[31]

The SDI became the 1980s equivalent of the 1970s ABM issue. In both instances the administration argued the advantages of a bargaining chip strategy, and the scientific community protested that the system would not work and would destabilize the arms race. But there was a difference; the Reagan White House did not have to contend with scientific critics on its own staff. Keyworth, whose background was as a weapons scientist, knew where he stood and argued forcefully for the SDI inside of government and in public debate. His weapons expertise helped bring the White House science office back into the national security field.

Science and Presidential Policymaking: The Bush Administration

The fortunes of the White House science office appear to have improved substantially in the Bush presidency. After an interim period with Reagan science adviser William Graham serving in a holdover capacity, President Bush chose Yale physicist D. Allan Bromley as his science adviser. Bromley, a distinguished scientist with a long record of professional service on advisory panels, was elevated in status from special assistant to assistant to the president for science and technology. The move was intended to symbolize that the science adviser's status was to be close to that of the national security and domestic policy advisers and a small number of other trusted aides.

Bromley recommended to the president in May 1989 the creation of the President's Council of Advisers on Science and Technology (PCAST).[32] Bromley also sought to breathe new life into the interagency coordinating machinery on scientific matters. The Federal Coordinating Council on Science, Engineering, and Technology

(FCCSET, pronounced "Fix-it" in Washingtonese) was given a broader mandate. Its meetings were chaired by Bromley, normally with senior officials (often cabinet members) from the agencies attending.

The FCCSET under Bromley has attempted to plan for the coordinated administration of crosscutting government programs. On matters such as the environment, new materials research, and competitiveness, agencies have given up a degree of program autonomy in the interest of a more effective governmentwide approach. While this approach is not wholly novel, Bromley has apparently succeeded in making interagency coordination a more effective working reality and as such has made a notable contribution to modern public management.[33]

Bromley acquired a larger staff than the office had under Reagan (in 1991 the staff included some forty people in professional and support categories) to carry out the wider range of duties. The science office appeared to have a level of activity and general impact in White House operations unknown in recent years. President Bush evidently took a substantial interest in the PCAST, meeting personally with the committee at its monthly meetings (at least from its establishment in February 1990 up to the fall of 1991).

The FACA and especially conflict-of-interest regulations delayed the PCAST's creation. The initial PSAC took two weeks to establish following President Eisenhower's decision to go forward with the idea. The Bush administration's PCAST required nine months of strenuous efforts from the date of the president's approval in May 1989 to the council's formal establishment in February 1990. The delay resulted in part from the need to obtain clearances, draft an extensive order, and obtain and file the proper charter. The principal reason for the delay, however, was steering a course through the complex conflict-of-interest requirements.

The initial opposition of the White House counsel's office, fearful of potential conflicts, was a major stumbling block. The president's legal advisers, sensitive to the charge that the Reagan years had been characterized by ethical laxity in high places, determined that the president must be protected at all costs from the appearance of conflicts. Industrial executives and academic leaders would, it was feared, have ties to the federal government that would arouse hostile press inquiries.

The committee moved through the legal labyrinth after intense debate over the financial disclosures required and the potential criminal liability of committee members for conflict violations. The members of the proposed council were given individual waivers immunizing them in part against criminal liability for advice rendered while serving on the council. The council met for the first time at Camp David in February 1990, and President Bush participated in the discussions for an entire morning.

The science office, despite its renewed vigor and despite the earlier Keyworth involvement with the SDI, still had no clearly defined role with respect to national security issues at the midpoint of President Bush's first term of office. It remained on the periphery of the internal debate of defense issues, even though the national security and science advisers were on cordial personal terms. At the beginning of 1991, as an augury of possible change, Bromley created a national security panel led by PCAST member S. J. Buchsbaum and Defense Science Board chairman John Foster. This marked potentially the most important involvement of the White House science office in national security affairs since the early days. The issues facing the panel were in some respects similar to the defense organization debates of the late 1950s: What should be the technical base of a new downsized military establishment? How should the Defense Department organize itself to relate technology to strategy and force posture in the new era?

In September 1990 the Office of Science and Technology Policy announced a technology policy for the administration, stressing the importance of "generic technologies."[34] To implement the technology policy goals, a $5 million appropriation for a "critical technologies" institute was included as part of the president's budget agreement with Congress, reached at the summit talks at Andrews Air Force Base in the fall of 1990. This institute was given its $5 million annual appropriation to review and harmonize the Commerce and Defense Departments' critical technologies lists and to advise on federal, state, and private sector roles in commercializing the technologies deemed "critical." A kind of federal contract research center or think tank attached to the OSTP was envisaged. In the summer of 1991, however, the plan collapsed, and the administration returned the funds to Congress (on the grounds that the proposal was an "industrial policy" scheme and that operational activity was inappropriate at the White House).

Despite the enhanced status for science advice under President Bush, the full postwar record suggests the fragility of an effective science advisory role at the presidential level. Built-in jurisdictional conflicts with other presidential advisers and staff units plus the vagaries of life at court render precarious the official existence of any adviser. The absence of a strong political base makes the science adviser especially vulnerable. But the full-time adviser at least is accountable and clearly part of the official family. The usefulness of the science adviser and his staff is generally more evident than that of the part-time advisory committee.

Part-time White House committees in general divide into two categories, the temporary commission and the standing committee. Both raise questions of reconciling the mores of scientific inquiry with the constraints of politics. Scientists, accustomed to the free exchange of ideas in an open environment, tend to view the disciplines of the policy process as political interference and bureaucratic heavy-handedness. They do not like to hear that presidents value loyalty as much as competence. Objectivity, professional repute, and other virtues taken for granted in science are of course important. But to the policymaker, these usually are not decisive. The adviser who enjoys a close personal or professional relationship with the president, who works in a timely fashion on matters about which the president has sought advice, and who performs advisory duties in confidence is most apt to be influential, whatever his or her professional standing.

A temporary committee can usually be tolerated more easily than a permanent committee. The former can normally attract a prestigious membership. A temporary committee exists in part to symbolize a commitment to action by the administration; its mere existence serves a purpose. It may be as much an example of the president speaking to the public as of experts or others speaking to the president. In cases in which the president did not know what he wanted from a commission or was surprised by the outcome, he usually quickly distanced himself from the effort. President Nixon's Commission on Campus Unrest and President's Johnson's Commission on Civil Disorders are examples of commissions quietly ignored.[35] President Reagan's Commission on Competitiveness was another such instance; the president was lukewarm toward some of its findings which appeared to be an endorsement of industrial policy.[36]

The AIDS Commission, though it settled down after a rocky start and produced a comprehensive plan that was widely admired, did not please the Reagan administration because its proposals went well beyond what the administration was prepared to accept.[37] The Commission on the Budget Deficit called for in the Gramm-Rudman-Hollings Act of 1987 failed even to produce an outcome. President Bush did not wish to use this vehicle because he did not want to pursue the deficit problem at the outset of his administration. The commission failed also because it received an unfavorable court ruling which prevented it from arranging a schedule of some closed meetings along with open sessions to conduct its business.[38]

Recent successful temporary commissions include the commission chaired by Alan Greenspan that produced an agreement on the financing of the social security system[39] and the Augustine committee on space. Earlier presidential commissions that were credited with achieving results include the 1965 Commission on Law Enforcement and Administration of Justice (generally credited with having prompted the Safe Streets Act of 1968) and the 1970 Commission on Obscenity and Pornography (its recommended postal regulation was enacted but its more radical reforms were ignored).[40]

The most robust period for the temporary presidential commission seems to have been the 1960s and early 1970s, roughly from the Warren Commission on the Assassination of President Kennedy to the passage of the FACA. The FACA seems to have slowed but not eliminated the creation of presidential commissions.[41] Presidents have become more cautious in creating them.

In 1990 there were some forty-five commissions or other outside advisory entities serving the president on a wide range of subjects.[42] Most of these were not in fact serving the president directly. They were working with particular departments. Some few served constituent units of the Executive Office of the President. Many were created by statute, and Congress in such cases participated in the designation of members. Applying the adjective *presidential* to an advisory commission is apparently often an effort to heighten the status of a departmental policy review effort.

Before the passage of the FACA in 1972, there may have been a somewhat larger number of presidential advisory commissions or other advisory bodies that either served the president or some part of the executive office (almost 200 in 1970, although many of them were probably not active). Most were created by executive order rather than by statute. Congress did not at the time participate in the naming of committee members as it has done more recently.[43] The president or his aides apparently were perhaps more inclined to create and use advisory bodies in the pre-FACA and pre-conflict-of-interest eras. Presidents may also find the prospect of having to share powers with Congress when they create a commission less appealing than the former practice.

A standing committee operates according to different rules. It has to adjust to the style of presidential decisionmaking. This will usually involve circumspection in public statements and harmony with other presidential aides as well as conformity to the administration's values and objectives. No matter how a science advisory system is structured, unwanted or unsolicited advice will always go unheeded. Presidents, like everyone else, seek opinions and information only on issues that concern them and only from people they trust (and often knowing in advance what they will be told).

There were, as of the time of this writing, fewer standing advisory committees serving the White House than temporary commissions. PCAST was the most prominent. The National Advisory Committee on Semiconductors, the Aeronautical Policy Review Committee, and the National Advisory Committee on Superconductivity also served the president's science adviser and the OSTP. The Information Technology Advisory Committee serves the OMB. The Office of the United States Trade Representative, a part of the Executive Office of the President, made use of a number of advisory committees reflecting corporate viewpoints.

Specifically serving the president were two additional advisory committees, the Intelligence Oversight Board and the President's Foreign Intelligence Advisory Board (PFIAB). These boards have operated in an entirely closed fashion and have been excepted from the FACA. The Intelligence Oversight Board was created by President Gerald Ford in 1976 to review and report to him on the intelligence community's activities. The PFIAB was established by President Eisenhower in 1956 to act as his informal board of consultants on a wide range of foreign policy and intelligence issues. It has had an off-again, on-again existence since.[44] Dissolved by President Carter, it was reconstituted in 1981 by President Reagan with fourteen members from outside of government. President Bush reorganized and recast the PFIAB into a group of five members chaired initially by the late senator John Tower. The Federal Employees Pay Comparability Act of 1990 added a Federal Pay Council with a statutorily defined membership to report to the president on pay adjustments.[45]

Conclusions

If the qualities that make for success in the science adviser role sound like those of a successful lawyer, economist, or other policy adviser, it is no coincidence. A science adviser is subject to the same code of behavior that applies to all presidential advisers. In each case the useful adviser is the one who is able to apply specialized expertise in the broad decisionmaking context that confronts the president. As seen in the discussion of the Nixon period, the substance of any White House policy matter is invariably blended with the politics of the issue, which has its own daunting complexity.

Even when an adviser is in a position to give, and the president is predisposed to receive, sound scientific advice, its relevance to and impact on policy are difficult to predict. For one thing science normally speaks with many voices; despite their commitment to objectivity, scientists rarely agree when the issue is broadened beyond the narrow confines of their disciplines (even on many purely technical matters scientists vigorously disagree).

The lack of agreement on broad issues has been true throughout the postwar period no matter what the issue: the wisdom of developing the hydrogen bomb, the safety of civilian nuclear reactors, ways of preventing acid rain, how to respond to global warming, coping with AIDS, and many others.

The recommendations of advisers must be weighed against a background of conflicting goals, intersecting events, and trade-offs among incommensurable values and tactical objectives. Social, political, and economic factors overshadow the narrowly technical aspects of science-related problems not because politicians, special-interest groups, and others meddle in issues about which they lack technical expertise. Technical factors are overshadowed because in matters of public policy there simply are no purely technical issues.

Moreover, science now pervades society to an extent that could not have been imagined forty years ago. The Defense Department, NASA, the Department of Energy, and the NSF, as well as outside groups, continually propose scientific initiatives, including the SDI, the space station, the superconducting supercollider, and cataloguing the human genome. There are now thousands of individuals in federal agencies and on congressional staffs who work on science-related problems or who are in some substantial way associated with them as proponents, critics, or analysts. Consequently, many more people than ever before are conversant with and sensitive to the scientific aspects of policy and often are better equipped than scientists to comment on the broader context in which such issues are debated.

The first and in some ways most urgent need for the nation with respect to science advice is to make sure that the agencies organize, manage, and employ their scientific resources effectively. The disproportionate attention given to science advice at the presidential level misses this fundamental point. Whether science advice is effectively integrated into policymaking—indeed, whether the federal government works or does not work well—is principally determined by what happens in the agencies. The White House science adviser cannot do the work of the individual agencies or supply the expertise they need.

Yet presidents will want their own advisers for the reasons already discussed: they will not trust the agency spin on technical or other issues; they will want to check an agency perspective with that of a trusted and experienced aide. This means the president's science adviser will be broadly experienced and have been with or known the president for a long time. He or she will not advise the president on science but on policy.

The science adviser can do what other presidential aides do to encourage or restrain agency action, provide a friendly ear and leverage the plans of agency officials when they conform to administration goals. These are important functions and naturally attract public attention. Yet the drama of high politics too often can obscure fundamentals that are evident at the operating divisions of government.

The factors that make for success in the advisory role in the White House are extensions of what we have seen at the agency level: breadth of vision rather than narrow expertise, and the ability to understand the needs of the policymaker. To these factors must be added a sensitive awareness of the climate of policymaking in the special setting of the presidency. The president must want to be advised, and the adviser must sense the timing, special circumstances, and nuances of the situation. As with the agencies and even more emphatically at the White House level, the advisory process should not be constrained by statutory mandates and specifications. Efforts to prescribe to whom the president must listen will usually fail and merely clutter the policy process. Flexibility and informality rather than legal mandate make for trust and confidence between adviser and decisionmaker. Arnold Meltsner wisely sums up this aspect of the advisory process:

It should not surprise us to hear that rulers will circumvent an advisory structure *or* reduce it to a minor role when that structure does not provide what the rulers want. Rulers may have to go through the motions, to give lip-service to the structure, but there is no way to box the rulers in. They do not have to listen to the advice of a particular advisory structure.[46]

Two common sources of proposals for specific advisory structures are congressmen and scholars of public policy, each for their own reasons.

Congressional efforts to prescribe to whom, how, and when executive officials and particularly the president must turn for advice are usually extensions of the battles over the substantive ends of policy. Failing to achieve the desired outcome, Congress may create a special unit to advise the president, such as the National Advisory Council on Oceanography. The idea was an attempt to maneuver President Johnson toward a NASA-type oceans agency.[47] The means was to redefine the question in such a way as to produce a new outcome. The effort predictably failed.

The other source of proposals to systematize the advisory process is scholars who seek to remedy the "inherent defects" or the "impediments" in the decisionmaking process. Yehezkel Dror, for example, has analyzed the defects cogently and proposed processes that would lead to more rational decisionmaking.[48] Alexander George has shown how systematic biases can mislead decisionmakers and distort the policy process in foreign policy decisions.[49] Irving Janis has a list of defects in decisionmaking.[50] While policymakers could greatly profit from these studies, in my experience it is the rare politician who enters office having read the literature of public policy. Once in power, policymakers become dominated by the oral culture. Almost no politicians read anything, let alone the scholarly literature, and virtually none could be persuaded to alter his or her style of decisionmaking. The pressures of presidential decisionmaking, the high-voltage and fast pace of events, and the almost daily crises cause presidents and their major advisers to be preoccupied with problems rather than with process and make it unlikely that they will pay serious attention to how they make decisions.

Scientists have a *déformation professionelle* of their own. Scientists feel comfortable in defining the issue so that it resembles a peer review process. But useful policy advice is not the same as advice on research objectives. Agency goals must be overlaid on the internal logic of research programs.

In the case of the president, who has no operating programs, the policy task is always one of defining ends and manipulating the symbols of office to dramatize those ends. The choice of advisers and the public face to be put on actions reflect the essence of presidential leadership. The president will find a way to avoid being constrained by congressional efforts to impose on him particular structures or sources of advice (especially in the current era of divided government). It is unrealistic to believe that any administration's priorities will be significantly influenced by different advisers or advisory structures.

Any president will frequently find uses for the temporary advisory committee. Rarely will the work of such a committee or commission be wholly original or heterodox (if it is sharply at odds with administration policy it will be ignored). The administration will communicate to the public through the committee just as the public or specialized publics communicate to the administration. A presidential commission is part of the pageantry and drama of democracy. It is America's version of the House of Lords.

Whether a standing advisory committee will be useful depends on the president. If the presidents and the presidential science adviser find such a mechanism useful, they should have it. If they do not and if they feel more comfortable without it, they should organize in any fashion they choose.

Science advice will make its way into presidential decisionmaking one way or another because of the inescapable complexity of public affairs. Neither the technocratic model of decisionmaking, in which scientists attempt to force policy debate into a peer review process, nor the model that seeks to dismiss the scientific expert as irrelevant to the debate over policy ends is appropriate for American democracy. The more realistic model involves a mix of expert advice and political judgment, of generalists and specialists. The boundaries between scientific means and political ends are subject to constant debate. Compromise and working accommodations between the advisers and the decisionmakers will be at the center of the process.

9 ‖ Science Advice and American Democracy

THIS BOOK has reviewed the experiences of science advisers in five agencies and in the White House, has described broadly how advisers participate in the policy process and how they contribute (or fail to contribute) to the workings of American democracy, and has sought to infer lessons on what contributes to a useful advisory role. The success achieved by the scientific advisory committees analyzed in this study can usually be traced to several key factors.

First, the committee should have a clear charter or mandate relating to an issue or problem of importance to the agency, to a major subunit, or to the nation. Most people manage to convince themselves of the importance of any committee they serve on, but there is no substitute for the reality of genuinely rendering service to your country on a matter of importance. Advisers involved in the decision to build the H-bomb, to develop the ICBM, and to plan the early space program clearly illustrate this point. These advisory efforts had the most important attribute of all: a sense of urgency resulting from addressing an issue of high importance to the nation.

Second, an identifiable client or point of access to the agency should exist, preferably with the agency official(s) in a position to act on the advice rendered. The official should genuinely want to be advised, should see a reason for having the committee (and should not create one unless there is a good reason), should be willing to devote time to the committee and to support it once it is created, and should make clear to subordinates that he or she will listen to the advice even if not necessarily accepting the advisers' recommendations.

Third, the committee should have an active chairman who blends such qualities as toughness, diplomacy, good timing, and assertiveness, and who manages effectively a group of potentially strong personalities. It is also helpful that the chairman have a long-standing friendship with the agency head or other senior officials, and it is indispensable that there be a relative harmony of outlook and shared values between advisers and decisionmakers. This means in practice that the chairman of the committee and the decisionmaker should be able to work along mutually agreeable lines. One person I interviewed put the point in these succinct terms: "The adviser and the policy-maker must be on the same political wavelength. If they share broad goals and agree on what they're trying to accomplish, the committee can be a very useful device. If they disagree politically, this is a recipe for disaster." Without the ability to discuss sensitive matters infor-mally and confidentially with the committee chairman, senior officials tend to adopt a perfunctory approach or are hostile to the advisory apparatus.

The chairman's role is so critical in part because of the almost ex-clusively oral tradition that operates in the higher reaches of the fed-eral government. Policymakers in general read almost nothing beyond the short summaries and briefing papers prepared by staff. They de-rive their impression of the adviser's message from what the chairman tells them or from oral updates of the panel's progress given them by staff. Occasionally if the committee's work attracts press or media attention, the policymaker may catch a sound bite summary of the findings on the evening news. Thus the personal interaction between the official and the advisers—most commonly the chairman of the formal committee—remains the critical variable.

Science advisers who are effective therefore almost always operate in the pragmatic rationalist mode. They bring to the advisory task or soon acquire a subtle understanding of how their efforts fit into the work of the agency or decisionmaker they seek to advise. They may choose to write a report using the rhetoric of the utopian rationalist, but they will almost always have subtly negotiated the terms of what they will say so as to mesh with the goals of their clients. This should surprise or shock no one, for all participants in the advising process share the larger purposes sought by the agency (or the agency would not have appointed them).

Fourth, there should be a well-chosen and fairly balanced commit-tee, blending some diversity of outlook along with congruence of pur-pose. A group too diverse cannot function, especially if there are genuine differences in philosophic orientation or basic values. Advisory com-mittees often have a dissenting or outlying member or members; this appears to ensure balance and vigorous debate. But appointing offi-cials, often advised by the chairman they have selected, usually take care to pick dissenters who ultimately share the agency's values. When they miscalculate and when advisers fall into irreconcilable internal differences, the advising committee produces a non-result, a bland or equivocal report or no report at all.

Finally, there should be adequate supporting resources in the form of staff time, travel allowance, and so forth, on the agency's part and commitment of time on the adviser's part.

Future Challenges

Although science advisory committees have sometimes played significant policy roles, the advisory system seems peculiarly vulnerable to neglect, misuse, or atrophy. Committees face an increasingly burdensome climate filled with numerous obstacles to successful functioning. Yet as the nation moves toward a more complex system of governance and as science and technology penetrate into nearly every aspect of government, the need for exposing the thinking of decision-makers to technically informed and experienced advisers seems more important than ever. The rationale that government will benefit from the advice and assistance of outside groups, as proclaimed in the language of the Federal Advisory Committee Act and affirmed by virtually every review of the subject, holds with equal force today. Moreover, agencies will apparently continue to create advisory committees whatever the legal regime set up to govern the process.

The challenge that the nation faces is to make the advisory system contribute to the larger ends of effective government rather than merely to constitute a layer of bureaucratic clutter or to prolong and complicate political controversies. Can the need for advice be met in the future under the legal regime and the informal understandings established to regulate advisory activity? Are there specific problems that should be addressed in the future relationship of science advisers and their agencies? The following sections discuss some of the complexities in the relationship between the expert and the policymaker so as to deepen our understanding of democratic governance. As James Allen Smith aptly observes, "The underlying problem—that of linking knowledge and power in an open society—does not present itself in the convenient form of a procedural flaw to be corrected, a structural defect to be repaired, or a disease to be prescribed for and cured."[1] The essence of self-government indeed lies in the continuing struggle to understand how knowledge and power interrelate and coexist, and how the citizen, elected and appointive officials, and scientists/experts play their parts in the political process.

The Meaning of "Fair Balance" in the Advisory System

Advisory committees struggle with the problem of ensuring that their membership is both representative and balanced. The major complaint of critics is that a narrow group of citizens, reflecting only partial interests, gets privileged access to decisionmakers under the guise of neutral expertise. Indeed, there is a tendency for advisers to slide toward clublike amiability and hew to an established line. The advisers must fight to stay intellectually alive and to find the level of creative tension that will nurture sharp debate and hard effort even if one cannot realistically expect committees to contend with or bridge fundamental philosophical differences.

At the same time that one decries the stodgy old-boy network, the experience of successful advisory committees indicates that broad membership and frequent rotation of members may work against a close working relationship between agency and board. Effective service on the Environmental Protection Agency and Defense Department advisory committees demonstrates, for example, that there is no substitute for experience with the intricacies and unique circumstances of an agency. The EPA originally wanted short terms of service on the Science Advisory Board but found that continuity in membership and accumulated wisdom in dealing with regulatory science would better serve its needs. The Defense Department had a similar experience, as did NASA and the Department of Energy.

What appears to critics as an entrenched old-boy network is primarily a manifestation of the complexity of modern government. The adviser must learn how to interact effectively with the agency, its clientele groups, congressional committees, media specialists, and the wider interested public. As a scientist gains experience with the agency's problems, he or she tends to be called on more and more by the agency.

For an adviser, part of learning the ropes is to practice restraint and to conform to the timing and style of the agency decisionmaking process. Scientists have special problems in this respect, and need to guard against the polar dangers of arrogance and irrelevance. The case studies show that scientists have trouble in distinguishing between a peer review and a policy advisory role. The natural inclination of scientists is to function as scientists, to evaluate the progress of their field, and to recommend support for meritorious research as defined by its contribution to the advance of knowledge.

Yet the "professional" adviser who makes a heavy commitment of time and psychic energy to advisory duties may become so much a part of the scenery that his or her usefulness is eroded. Advisory duties must be challenging enough to engage the individual's interests, time, and vital energies. But the effective adviser should avoid becoming in effect a full-time official. The adviser therefore faces a paradox: he or she must become a true insider to accomplish anything; but in doing so the adviser may lose the fresh view, detachment, and outsider qualities that are urgently required.

What about political balance? Given the fact that the most effective advisers must be politically savvy, how much of a role should politics play in the appointment process? There is a constant pressure for patronage appointments on advisory boards, especially the boards with some degree of prestige and public visibility.

Depositing persons with influential backers onto an advisory committee made up of scientific luminaries and other notables is one harmless way for a politician to pay a debt. Sprinkling individuals with political connections through the advisory ranks may also be a small price to pay for a flourishing democracy. Indeed, having a committee chairman who has close personal relationships with the cabinet secretary or other senior officer can be a vital factor in the committee's success. Patronage considerations therefore should not be universally condemned. All this said, one still wonders whether the nation is well served by the application of the same kinds of political tests to advisory board membership as apply to high-level appointive positions within the government.[2]

It is difficult to document the point, but there appears to have been a gradual increase in the number of patronage appointments in some areas where technical criteria used to predominate. This is in part simply the reflection of greater White House involvement in appointments of all kinds. The pressure for more attention to political criteria in advisory appointments no doubt also reflects the increasingly partisan tone in American politics that has prevailed since the Watergate and Vietnam eras.

Equal opportunity and affirmative action are another aspect of the balance problem. Attempts to promote equal opportunity sometimes lead agencies to depart from strict adherence to technical criteria on appointments to science advisory committees (though a systematic search can usually find qualified minority physicians, engineers, researchers, and people with other skills). The broader policy-oriented advisory boards, like corporate boards, will have no trouble drawing from a wide pool of talent.

Finding the right balance among legitimate but conflicting criteria in the composition of advisory committees will never be easy. No simple recipe or formula exists. The choice of individuals to serve, and in particular the selection of those who will occupy the leadership positions within the advisory structure, are some of the most important decisions that an administrator or policymaker can make. An agency head should take seriously the health of the advisory committees that serve the agency. This means both training advisers so that they can be useful and keeping them intellectually engaged. More time spent in searching for candidates and, once selected, making them effective members of the committee and of the agency's team will yield important dividends.

Many committees simply recycle their roster of advisers without a conscientious search for new talent. The fact that it is difficult to find the right combination of technical capacity, willingness to serve, and practical experience makes it all the more important to spend time in the recruitment and training of members. In the end, managing an advisory committee effectively is akin to good management generally: formulate the objective, assemble the team, motivate the members, and evaluate the results.

Should the Federal Advisory Committee Act Be Amended?

Should new legislation be enacted to amend the 1972 advisory committee act, perhaps as S. 444 tried to do in the summer of 1990? In my view, the needs of advisory committees are so diverse that statutory definition of specific skills, membership categories, and uniform procedures would be impractical and unwise. The agencies are the best judges of who should serve. The loose requirement for fair balance contained in the present legislation is sufficient. But a few changes in the FACA might be useful to clear away misunderstandings and help the federal government to make more effective use of science advisers and other sources of outside expertise. Some of its provisions have been ignored, and very few people know or care much about the FACA in Congress or at senior levels in the executive branch. The country would in any case benefit from a thorough airing and debate of the issues.

Such a debate might well lead Congress to amend the FACA to remove some of its burdensome features and to bring the theory of open meetings into harmony with actual practice and administrative reality. The principles that should guide the debate and possible revision of the law should include an affirmation of the broad commitment to open meetings and open processes of government contained in the statute; clearly, however, the closing of meetings should be allowed especially for preliminary, working group, or informal discussions. It serves no purpose to continue the pretense that confidential deliberations injure democracy. If agencies abuse the privilege, stringent oversight can remedy the situation. Oversight hearings (practically nonexistent since the death of Senator Lee Metcalf) can be held to call attention to significant abuses or failure to comply with the spirit and letter of the act.

There is also a need to change the law to simplify the reporting requirements for agencies when they create advisory committees. Little purpose is served by filing charters with congressional committees or providing a rationale for closing a meeting to the Library of Congress. The two-year charter renewal provision should probably be changed to a three-year requirement. This three-year limit would be more likely to be taken seriously and enforced.

The burden for creating committees should rest primarily with the agencies, as now, after "consultation" with the Office of Management and Budget. But the roles of the OMB and the General Services Administration need to be clarified. The OMB should be mainly an adviser to the GSA and to the agency, not a policeman.

In addition, the creation of specific advisory committees by statute, which has become an increasingly common practice, is not consistent with the spirit of the advisory committee act. The FACA sought to establish a process to regulate the creation and use of advisory committees *by the agencies*. It did not envisage that Congress would create specific advisory committees and attempt to delineate their roles in agency decisionmaking. The experience of advisory committees that are mandated and imposed on the agency is not encouraging. Inflexibility and bureaucratic formalism often result from congressional intervention of this kind, especially when Congress attempts to specify the timing and categories of decisions requiring committee involvement.

A series of technical amendments contained in the proposed S.444 deserve serious consideration as do several minor amendments suggested by agencies. For example, the definitions of *committee* and *subcommittee* as used in the FACA are in need of clarification. State and local government officials who are acting in their official capacities should be exempt from the application of the advisory committee act. The meaning of *consult* with respect to the GSA and OMB roles should be clarified so as to vest final authority in the agency head to establish, manage, and terminate committees. The president, on the recommendation of an agency head, should be allowed to terminate statutorily mandated advisory committees (sometimes called deadwood committees) that are no longer needed.

When it enacted the FACA, Congress did not envisage creating through statute a large number of specific advisory committees and imposing them on agencies or the president. The FACA sought to change the way the executive branch operated when it turned to experts for advice. Senator Metcalf, Representative John S. Monagan, and their colleagues did not think they knew better than the executive agencies when to create an advisory committee. Congress in the future should remember this wisdom. Specifying when a committee should exist, whom it should represent, and what it should do is the responsibility of the executive branch. The executive branch should be able to create and abolish committees when and as appropriate, provided that the process is transparent and in conformance with the principles laid out in the FACA.

The OMB in implementing Reorganization Plan No. 1 in 1978 transferred the committee management function to the GSA along with most of the staff members who previously administered the system. But it did not surrender actual authority, as is evident from the many occasions when the OMB, through its consultation process, attempted to block the formation of new committees. This anomaly has been a source of irritation and confusion in the agencies for more than a decade.

Since the mid-1980s the GSA Committee Management Secretariat has done a workmanlike job of administering the advisory committee system. In the early 1980s it overzealously interfered with the agencies, creating a near revolt on one occasion on the part of the agency officials administering the system. Since the mid-1980s the GSA has conceived of its role as essentially that of a record keeper and cheer-leader rather than a policeman and inspector. However well the GSA has performed its role recently, it needs input from the OMB to handle the function. The GSA does not have enough knowledge of the agencies to give useful advice. It should normally seek comment from the OMB budget examiner as a part of its review process. The OMB should respond promptly to requests for comment so that the whole process of consultation comprises no more than two or three weeks. The GSA Secretariat needs an increase in staff to ensure prompt processing of requests and smooth administration of the system.

The system should be modeled on how it operated in the period from 1975 to 1977. This period was probably the most successful phase in the administration of the advisory committee system. Agencies would file a charter with the committee management office in the OMB (in keeping with the requirement to consult). The agency would be promptly informed of the OMB's views. On those occasions when an advisory committee wanted by an agency head could embarrass the president, the OMB acted as a gatekeeper to safeguard presidential prerogatives. There was a uniformity and overall discipline to the process by virtue of having a senior full-time official in charge so that individual budget examiners could not exercise arbitrary control.

The GSA and the OMB offices should assist the agencies in devising their financial disclosure and other conflict-of-interest provisions. Significant steps could be taken by agreement among the Office of Government Ethics, the Justice Department, and the OMB to improve the process of compliance with conflict-of-interest laws and regulations, issuance of waivers, and filing disclosure statements that would protect the public interest without unnecessarily burdening advisory committee members.

Conflicts of Interest

The major conflicts of interest covered under current and pending legislation largely refer to those instances in which insiders (that is, full-time government employees) seek to exploit their special position by gaining lucrative positions in the private sector after their government service or else seek to exploit their previous position once they are in the private sector. Current laws cover only indirectly situations in which outsiders seek to advance the special interests of their firms or employers by exploiting access to inside information gained through their advisory service.

The government must of course protect the public interest against both kinds of abuses, by insiders and by part-timers serving on advisory committees. Conflicts regarding science advisers are an especially sensitive issue in the regulatory arena. But conflicts of interest in the narrow sense are a vastly overstated danger for most advisory committees. The group dynamics of committees of distinguished citizens militate against any member seeking to influence government policy to advance narrow personal or institutional interests. The clash of different interests and perspectives ensures that no single narrow interest will dominate a committee's deliberations. Moreover, the concept behind the FACA is that the public interest is protected through competition among points of view and that the decisionmakers benefit from the interchange and know when and whether to accept advice from outsiders.

The government should not of course be beholden to, or overly dependent on, any group of advisers. Ever more stringent rules and regulations do not, however, contribute to the goal of protecting the public interest. Rather, the search for the last ounce of protection by assailing the advising apparatus all too often is a sideshow that merely confuses the public and feeds the populist illusion that all government is corrupt. The inner check of professionalism rather than the legislature acting as external policeman is the better route to high ethical standards and good performance in the executive branch.

Current laws that focus on the postemployment activities of government employees have perhaps reached the point of overkill, particularly as applied to high-level appointees. It would be folly to apply the same overkill to the special circumstances of advisers who serve only intermittently. The nation has already added unnecessary burdens to service on advisory committees without full or explicit recognition that this has been happening and with little or no added protection of the public interest.

There is no simple way to counter the populist distrust of the expert so common in our political culture and so easy to exploit amid charges of wrongdoing. Balancing experts with public interest representatives in some formalistic fashion on all committees is no solution. Public interest advocates, consumer representatives, advisers from the union movement, and many others may be useful participants in an advisory endeavor, especially if the problem relates broadly to their expertise. But the advisory membership should normally comprise individuals chosen for their broad experience and judgment, not interest group representatives.

The advisory system, by virtue of its openness and pluralism, is shot through with interested as well as disinterested advisers. Nowhere is the problem more difficult than in the case of science advice in the regulatory process. As one member of the EPA Science Advisory Board told me: "There is no one in this country who we might want on the board that does not have some conceivable conflict. . . . It is surely no answer, and it is not even imaginable, to place on this or similar boards someone with no connection whatever to the field."

This is not to say that improvements cannot be made in current conflict-of-interest practice. For example, a subtle dimension of the conflict problem is the appearance of bias or parochialism in the selection of science advisers. While it is true that agencies are almost never dependent on any single committee or single source of advice and are always able to discount advice they regard as biased, the appearance of bias should be avoided and can be avoided by careful selection of advisers. A committee studying the future of the defense technology base made up exclusively of defense industry executives, or a panel recommending increased funding for biomedical research composed exclusively of biomedical scientists, invites public ridicule and suspicion.

A partial answer in some circumstances might be to increase the numbers of advisers from universities and not-for-profit institutions on matters relating to industry. Conversely, industrialists carry more weight in commenting about the problems of colleges and universities than those who are directly involved in academic governance. Peer review panels are a special case and should be largely composed of the relevant scientific specialists. It is not realistic or appropriate, however, to limit in some blanket fashion appointments of science advisers from industry, because three-quarters of all scientists and engineers are employed in industry.

In addition, a useful step would be the adoption of uniform disclosure requirements modeled on the recommendations of the Administrative Conference of the United States for all advisers classed as special government employees and perhaps also for those considered *representative*. The members of advisory committees should not have to face the same standards as full-time government employees and should not be subject to potential criminal liability for serving on advisory panels. The distinction between members of representative advisory committees and those who are defined as special government employees is an anomaly as it is now interpreted because it imposes burdens equivalent to full-time employees on the special employees and no disclosure or other requirements on the representative members.

Finally, agencies can protect themselves against the appearance of conflicts and enhance the quality of the advice they receive by taking the following steps: searching for new advisers in a serious way and through a transparent process; avoiding overrepresentation of any single group; having explicit recusal policies and making such policies known to the public (while providing, as appropriate, for waivers to protect individual panel members); and talking about the problems of preventing conflicts at the time a committee first convenes and maintaining oversight of potential conflicts by the committee's executive secretary (following the longstanding practice of the National Academy of Sciences).

The ultimate solution to the problem of protecting the public interest is the intelligence and judgment shown by responsible officials, not efforts to strip officials of discretion.

Should Scientists Provide Technical Advice Only?

Frederick Seitz has observed that the "leaders of the National Academy of Sciences have learned over the years that their effectiveness is greatest when the organization is able to deal with topics of two types, namely, matters of an almost exclusively technical nature having great public interest and on which professional advice is widely welcome, or with issues of primarily professional interest such as those related to promising frontiers of basic research."[3] Seitz's formulation is close to Robert Wood's concept of scientists as an "apolitical elite" and Yaron Ezrahi's notion of the "utopian rationalist" science adviser.[4] Smith has also broadly analyzed the role of experts and analysts in operational matters and found their influence to be more readily apparent than in the value-laden and ambiguous realm of politics.[5]

These analyses no doubt accurately capture an important difference between the well-focused technical panel and the omnibus advisory committee with a rather diffuse mandate. The working groups or the

standing committees that operate with a clear assignment, by definition, can be more effective than those that struggle to define their task
and their jurisdiction. Often in the case studies in this book the more
focused subgroups operating under an overall parent committee seemed
to get more done than the parent body. They not only seemed to have
a clearer focus but were less subject to the requirements of the FACA.
They were able to meet informally provided that they reported to the
agency through the parent body, and some did not even need to become officially established. It is evidently the case that scientists work
more effectively at the agency level than at the White House level.
The line agencies engage after all in operational tasks requiring technical expertise and clearly need the advisers.

In short, it is useful to break down the advisory task into manageable parts. But having said all this in support of Seitz's observations,
the significant point is that scientists in the United States differ from
their counterparts in other nations in the degree to which they participate in broad policy decisions, not merely in policy *for* science decisions.

The advisory committee that deals with narrow technical or professional concerns is making an important contribution but is not doing
what is unique and significant about science advisers in the United
States. The paradox facing scientist advisers is that they are called on
to function in a setting where their traditional skills are inappropriate
or at least not well suited to the difficult and ambiguous tasks they
attempt to perform. They are called upon to blend the purely technical aspects of issues into a larger, more confused, and value-laden whole
where there are no precise answers. Scientists are called on because
they have special expertise, but they must think and act like politicians
if they are truly to be useful.

Agencies already have technical capacities within their own ranks.
The easy technical tasks can be handled by the agency's own staff.
The decisionmakers need the outside advice precisely because they
have difficulty making the partial perspectives blend with each other
and with the agency's or the administration's larger priorities. There
is no way for science advisers to escape the conceptual problems and
the confusions of politics if they are dealing with the agency's true
problems and are seeking to give truly useful recommendations.

Scientific advisers get the best results when they are not acting as
schoolmasters explaining the science of an issue, but when they attempt to make judgments and fit the issue within the total context of

the aspirations, constraints, and trade-offs facing the policymaker. For the expert to function only in a scientific capacity runs two opposite risks: the adviser may be irrelevant because the strictly scientific aspects of the problem are of limited interest; or conversely the adviser may incorrectly assume that the science explains the whole of the problem. Some of the complaints about the erosion of science's authority in the public policy process are based on this mistaken assumption that policy conclusions follow inexorably from the scientific facts.

Indeed, to the extent that they learn enough about the agency problem to be useful, scientists will function less and less as scientists and more and more like the politicians or generalist administrators they advise. The scientist adviser over time tends to suffer a degradation of his or her purely scientific capacities and standing so that, paradoxically, the truly active or productive scientist will rarely have the time or incentive to be an adviser.

Sheila Jasanoff's analysis of the function of scientific advice in regulatory policy has broad relevance for the scientist's role in policymaking generally. She rejects the technocratic model for regulatory decisionmaking, in which issues are defined exclusively in scientific terms (the "good science" approach, which conceals policy preferences beneath a seeming neutrality). She finds equally unacceptable the participatory politics paradigm, in which science is seen by some liberal critics as merely another elitist structure defending the establishment by pseudo-objective argument. The function of the scientific adviser is rather to engage in dialogue and consensus building with policymakers on both the policy ends and the scientific means of regulation.[6] Science alone cannot definitively resolve environmental controversies, but it can help to legitimate policy by defining the boundaries of the technically feasible and the politically acceptable. Moreover, science almost invariably plays a part in every major issue even if it seldom provides the definitive answers.

When the Founders substituted Madison's "new science of politics" for established monarchical or religious authority, they placed disputes over the respective boundaries of reason and of political compromise at the center of the governing process. Practical reason became a driving force for government action. The task of government was to devise concrete solutions to practical problems. But since experts disagree and their parochial solutions often conflict, the process of political accommodation and compromise was also built into

America's constitutional design. A useful shorthand description of America's pluralist democracy is as a triangular process: scientists espousing their partial versions of the truth, politicians brokering among those competing claims, and courts enforcing due process and procedural fairness.

Most science advisers interviewed for this study know that they must do more than offer technical expertise. They normally develop a close knowledge of the agency they serve and become familiar with the nontechnical dimensions of the problems on which they advise. But there remains the pretense that science advice is merely a matter of providing the facts. It is sometime argued that the public would be confused to discover that scientific truth is contingent, subject to dispute or interpretation, and not a source of ultimate authority. The public once again is wiser than the pundits. Americans generally have the balance and sense of proportion to recognize the strengths and weaknesses of an argument. They have a healthy respect for science but also a sober disinclination to accept as truth a scientific argument that conflicts with deep-rooted values or common sense.

The public's understanding of complex technical issues is related to the question of how technical issues should be analyzed in the policy context. It is sometimes held that the public cannot understand such issues as global warming, waste disposal, and strategic missile defense because of inadequate knowledge of the science of these issues. The solution under this view is to increase scientific literacy. A study done for the Charles F. Kettering Foundation in 1990 casts the problem in a different light.[7] The study, which focused on the public's understanding of global warming and the disposal of solid waste, showed that "the lack of scientific knowledge is not what blocks the public from thoughtfully considering most highly scientific issues. Far more important than facts and figures is a framework within which the issue can be assessed."[8]

The respondents interviewed for the Kettering study tended to deal intelligently with scientific uncertainty or conflicts among experts, by relating the uncertainty to some aspect of personal experience they could understand, and were prepared to revise their provisional judgments in the light of new evidence. The public's judgments about scientifically complex issues will usually be in accord with the views of scientists once the issue has been presented in a broadly intelligible framework. Public opposition to an unpopular policy option may not change, however, no matter how much technical information people receive. Finally, the reasons that people reject some solutions to technical problems favored by the experts "may involve other factors that are essentially nonscientific: . . . the reluctance to raise taxes (necessary for some proposals to curb greenhouse emissions), concern about property values (and the not-in-my-backyard . . . principle), and the history of an issue (many respondents felt that the history of nuclear power in the U.S. is replete with bad faith)."[9]

Conclusions

How in practical terms should the scientist be integrated into the policy process? How much independence should the advisers enjoy? The advisory committee must be in some sense independent to perform a useful function. If the advisers adapt so fully to the agency culture that they become merely an extension of it, agency problems will simply reappear in new form.

Yet there are clearly practical reasons that the advisers' independence cannot be absolute. Science advisers are given access to sensitive information and to the agency's inner policy deliberations. They should not treat this access casually, engage in unauthorized public disclosure, or otherwise behave as if access brought no constraints on their behavior. Policymakers will inevitably require a certain comfort level with their advisory panels or they will find myriad ways to block the advisers from meaningful participation in decisionmaking.

Subtle relationships arise in advisory systems with both internal committees and external advisory bodies (such as the National Academy of Sciences or the Rand Corporation). Overlapping memberships can complicate the rules of the game. Clearly, mutual trust is the prerequisite for achieving a satisfactory advising relationship. This cannot be demanded or presumed by either side. It must be earned through experience. American society has many institutions that play some role in the overall knowledge creep by which ideas seep into the public discourse and ultimately influence policy. The distinctive contribution of the "insider outsider"—those nongovernment scientists serving on official committees—is to act as broker between the larger world of ideas and the agency's specific needs and objectives. When they assume this role, advisers learn the delicate minuet of functioning as insiders with inevitable constraints without forfeiting their professional identities as outsiders.

Science advisers or committees lacking staff support will usually be incapable of performing their missions. To create a committee without a staff backup, as in the case of the State Department, serves little purpose. The other committees analyzed in this study have had in varying degree sufficient staff resources to do something useful. At a minimum the necessary elements seem to include the following: a full-time secretariat consisting of several persons; additional staff support, preferably drawn from the rest of the agency, for each separate task group established; an adequate budget, including some funds for travel, consultants, and contract research; and a career pattern within the agency that makes a support role to the advisory board a desirable assignment (both for promising younger officials and for senior people). The advisory board's secretariat should comprise persons with enough stature to permit easy interaction with other senior executives in the agency (both appointive and career) and with the committee members.

Careful thought must be given both to the target of the committee's advice (normally the secretary or deputy secretary or other senior official) and to the nature and timing of the communication process. The public report is only one step in a long process of interaction, and it may not be the most significant one. The most important result of the advisory work may simply be what is communicated to the policymaker informally, usually through the chairman. The public report can often be merely a formality. The oral culture of Washington has been noted above, and this perhaps explains why many committee reports are so bland and so inconsequential; the real message has been delivered and an impact has already been achieved.

The transparency of the process required under the FACA will, however, normally call for a written product that will be publicly disseminated. Publication may also serve to dramatize the study for agency staff and for constituency groups. But the aim is not authorship; it is advice. The advisory committee is providing advice to the client and normally not to the general public. The presidential commission is an exception; it is almost always speaking to the general public as well as to the president.

Various advisory committees have experimented with formal devices to require operating officials to consider the recommendations coming from the committee. The Defense Science Board and the EPA Science Advisory Board, for example, have had at various times a procedure whereby recommendations must receive a written response and a detailed account of actions taken by the appropriate operating division. Proposals have been made that the recommendations of presidential commissions require status reports from agencies on progress toward implementation of committee recommendations.

Are such formal procedures desirable? In general such devices work only if they reflect mutual trust between the advisory body and the agency. If a favorable working relationship exists, the formal requirement is unnecessary; if trust does not exist, the formal requirement cannot be effective.

The issue resembles the question whether agencies should be free to structure their advisory panels as they choose or whether statutory mandates should govern the process. Prescribing detailed administrative procedures by statute usually interferes with agency flexibility. Agencies will usually find a way to defeat or render impotent administrative forms imposed on them. Requirements to consult with advisory committees or to "consider" the advice may be meaningless.[10]

The desire to structure advice in some systematic way in order to nudge the administrative process in a desired direction was clearly one motive behind the adoption of the FACA. But the FACA seeks to create a general process to regulate how agencies receive advice, not to encourage congressional committees to specify what advice is given on what subjects and for what purposes.

America has evolved a loose-jointed system of government with a permeable outer skin. Ideas, interests, and people make their way back and forth across the boundary between public and private sectors with relative ease. Science advisers assembled into committees have found a niche in the overall system as the means whereby the wider energies of society are funneled to the decision points within government. At their best the outside scientists have played important roles in the overall scheme of things. They have helped to reformulate complex problems for decisionmakers and to revise the agency's strategies for tackling major problems.

While advisory committees perform a democratizing function as well as provide expertise, they cannot be participatory in the tradition of the radical critics of liberal democracy. The science advisory committee cannot and should not be transformed into a populist device to represent disaffected groups. Where this has been tried it has failed and has sown mistrust. To be truly useful scientists must be part of the political process. But they cannot be merely advocacy groups or an instrumentality to achieve popular participation in government.

Science advisers may from time to time sink into inaction or become another element of clutter in the nation's disorderly policy process. There will be a need to check the growth of advisory committees from time to time, eliminating the useless ones and revitalizing others. Americans should also seek to get their money's worth from the network of advisory committees. If the nation goes to the trouble of having a system of advisory committees, it should pay enough attention to their management and invest enough energies to make the system work at its best.

Practical attention to how science advisers and advisory committees operate may contribute modestly to the improvement of this aspect of public affairs. There is no uniform recipe for how this can be done, as agencies and advisers alike grapple with special circumstances and mandates for action. The organizational cultures in which advisers function will shape their styles of operation. Advisory committees will always be a challenge and sometimes an irritant to the decision-makers they serve. The test of their usefulness is whether the nation is better off with senior officials' having to expose their thinking to a body of knowledgeable and sympathetic outsiders. So long as we do not fall into the error of believing that advisers can produce simple answers to complex problems, the nation can profit from the creative energies that are mobilized for public purposes and take pride in the willingness of so many gifted citizens to serve their country.

Although policy formulation should properly rest with accountable officials, the experience of American democracy suggests that the modest check on executive authority exercised by advisory panels serves a useful purpose. Policy can never be solely the simple reflection of scientific expertise, no matter how broad gauged and well informed that expertise may be. Yet most major issues have technical dimensions of some kind or require the involvement of scientists at some stage. The policy process seems clearly bound to wrestle with the issue of how to blend the findings and methods of science with the power struggles and value conflicts of democracy. The challenge to the nation is to reconcile the integrity and the disciplined search for truth of science with the openness and procedural fairness of democracy.

Notes

Chapter One

1. At the end of fiscal year 1990 there were, 1,071 advisory committees of all kinds in the federal government. The number of advisory committees has apparently stabilized at about 1,000, for at least the past decade, with turnover of some 10 percent a year as new committees form and old ones disappear or atrophy. Committees like the government agencies discussed by Herbert Kaufman, in *Are Government Organizations Immortal?* (Brookings, 1976), may tend to live on in name even if they function with diminishing energy. But they do have a built-in "sunset" provision in that they are supposed to be formally rechartered at periodic intervals or else expire. Some committees are temporary in nature and cease upon completion of their assignments. The total membership on federal advisory committees in 1990 was 21,699 people (of whom 10 percent, or 2,107, were full-time federal employees); 3,720, or 17 percent, were from state and local government; 7,827, or 36 percent, from private industry; 6,886, or 31 percent, from the nonprofit sector; and miscellaneous 1,159, or 5 percent. *Nineteenth Annual Report of the President on Federal Advisory Committees, Fiscal Year 1990,* p. 6.

2. The GSA Committee Management Secretariat divides all advisory committees into the following categories: grant (or peer) review 227 (20 percent); scientific-technical (other than grant review) 308 (27 percent); nonscientific 260 (23 percent); national policy–issue orientation 246 (22 percent—perhaps a third or a half of these policy committees are explicitly involved with technical-political issues); regulatory negotiaton 5 (0.1 percent); and other 82 (7 percent). Ibid., p. 2.

3. Robert Wood, "Scientists and Politics: The Rise of an Apolitical Elite," in Robert Gilpin and Christopher Wright, eds., *Scientists and National Policy-making* (Columbia University Press, 1964), pp. 41–72.

4. Wallace S. Sayre, "Scientists and American Science Policy," in Gilpin and Wright, eds., *Scientists and National Policy-Making*.

5. For a discussion of the NAS, see Philip M. Boffey, *The Brain Bank of America: An Inquiry into the Politics of Science* (McGraw-Hill, 1975). See also Harold Orlans, *The Nonprofit Research Institute: Its Origin, Operation, Problems & Prospects* (Carnegie Foundation, 1972); Bruce L. R. Smith, "The Nongovernmental Policy Analysis Organization," *Public Administration Review*, vol. 37 (May–June 1977), pp. 253–58; Bruce L. R. Smith, *The Rand Corporation: Case Study of a Non-Profit Advisory Corporation* (Harvard University Press, 1966); and R. Kent Weaver, "The Changing World of Think Tanks," *PS: Political Science and Politics*, vol. 22 (September 1989), pp. 563–78.

6. For an illuminating discussion of the definition of the meaning of *advisory committee* under the FACA, and how the courts have interpreted the statute, see Michael H. Cardozo, "The Federal Advisory Committee Act in Operation," *Administrative Law Review*, vol. 33 (Winter 1981), pp. 1–62.

7. James Allen Smith, *The Idea Brokers: Think Tanks and the Rise of the New Policy Elite* (Free Press, 1991), p. 230.

8. See Harold F. Gosnell, "British Royal Commissions of Inquiry," *Political Science Quarterly*, vol. 49 (March 1934), pp. 84–118; and Hugh McDowall Clokie and J. William Robinson, *Royal Commissions of Inquiry: The Significance of Investigations in British Politics* (Stanford University Press, 1937).

9. Ronald Brickman, "Comparative Approaches to R&D Policy Coordination," *Policy Sciences*, vol. 11 (August 1979), pp. 73–91; and William Golden, ed., *Worldwide Science and Technology Advice to the Highest Levels of Government* (Pergamon, 1991).

10. That is, they have usually been limited to advising on the allocation of R&D resources, peer review evaluations, and the planning of research activities. See Ronald Brickman and Arie Rip, "Science Policy Councils in France, the Netherlands, and the United States," *Social Studies of Science*, vol. 9 (Summer 1979), pp. 167–98.

11. Stephen Krasner, " United States Commercial and Monetary Policy," in Peter Katzenstein, ed., *Between Power and Plenty* (University of Wisconsin Press, 1978), pp. 51–87; Peter Katzenstein, *Small States in World Markets: Industrial Policy in Europe* (Cornell University Press, 1985); and Phillipe Schmitter and Gerhard Lehmbruch, eds., *Trends toward Corporatist Intermediation* (Beverly Hills, Calif.: Sage Publications, 1979).

12. Mort Grant, "The Technology of Advisory Entities," *Public Policy*, vol. 10 (1960), p. 93, n. 1; and Don K. Price, *Government and Science* (New York University Press, 1954), chap. 5.

Chapter Two

1. See Yaron Ezrahi, *The Descent of Icarus: Science and the Transformation of Contemporary Democracy* (Harvard University Press, 1990), esp. chaps. 1, 4; Don K. Price, *America's Unwritten Constitution* (Louisiana State University, 1983); and Bruce L. R. Smith, *American Science Policy since World War II* (Brookings, 1990), chap. 1.

2. Judith N. Sklar, *Montesquieu* (Oxford University Press, 1987), pp. 120-22.

3. "A Letter of Luther Martin to the Citizens of Maryland" (1788), quoted in Ezrahi, *Descent of Icarus*, p. 108.

4. Edmund Burke, "A Letter to a Noble Lord," quoted in Ezrahi, *Descent of Icarus*, p. 119.

5. Edward Samuel Corwin, *The "Higher Law" Background of American Constitutional Law* (Great Seal Books, 1955).

6. Richard Hofstadter, *The Paranoid Style in American Politics and Other Essays* (Knopf, 1965), and *Anti-intellectualism in America* (Knopf, 1963).

7. Yaron Ezrahi, "Utopian Rationalism and Pragmatic Rationalism: The Political Context of Science Advice," *Minerva*, vol. 18 (Spring 1990), pp. 111-31.

8. Price, *America's Unwritten Constitution*.

9. David Flitner, Jr., *The Politics of Presidential Commissions: A Public Policy Perspective* (Transnational Publishers, 1986), pp. 7-8.

10. Washington's Sixth Annual Address, quoted in Carl Milton Marcy, *Presidential Commissions* (King's Crown Press, 1945), p. 109, fn. 4.

11. Richard Hofstadter, *The United States: The History of a Republic*, 2d ed. (Prentice-Hall, 1970), p. 205.

12. A. Hunter Dupree, *Science in the Federal Government: A History of Policies and Activities to 1940* (Harvard University Press, 1957), pp. 26-27.

13. Quoted in Marcy, *Presidential Commissions*, p. 8.

14. Ibid.

15. Bruce L. R. Smith, ed., *The Higher Civil Service in Europe and Canada: Lessons for the United States* (Brookings, 1984), chap. 1.

16. Robert V. Bruce, *The Launching of Modern American Science* (Knopf, 1987).

17. For the evolution of the American science and technology policy system up to 1940, see Dupree, *Science in the Federal Government*.

18. Peter J. Kuznick, *Beyond the Laboratory: Scientists as Political Activists in 1930's America* (University of Chicago Press, 1987).

19. For a detailed account of these developments, see Smith, *American Science Policy since World War II*, chap. 3.

20. Don K. Price, *Government and Science: Their Dynamic Relation in American Democracy* (New York University Press, 1954), chap. 2.

21. H. L. Nieburg, *In the Name of Science* (Chicago: Quadrangle Books, 1966).

22. Lester M. Salamon, "Rethinking Public Management: Third Party Government and the Changing Forms of Government Action," *Public Policy*, vol. 29 (Summer 1981), pp. 255-76. For a general discussion, see Harold Seidman and Robert Gilmour, *Politics, Position, and Power: From the Positive to the Regulatory State* (Oxford University Press, 1986), pp. 119-22.

23. Bureau of the Budget, *Report to the President on Government Contracting for Research and Development* (The Bell Report) (May 16, 1962), reprinted in *Systems Development and Management*, Hearings before a subcommittee of the House Committee on Government Operations, 87 Cong. 2 sess. (Government Printing Office, 1962), pt. 1, pp. 191-337; B. L. R. Smith, "The Nongovernmental Policy Analysis Organization," *Public Administration Review* (May-June 1977), pp. 253-58; and *Leadership for America: Rebuilding the Public Services*, Report of the National Commission on the Public Services, Paul A. Volcker, chairman (Washington, 1989).

24. Charles J. Hitch and Roland N. McKean, *The Economics of Defense in the Nuclear Age* (Harvard University Press, 1960).

25. See James Q. Wilson, *Bureaucracy: What Government Agencies Do and Why They Do It* (Basic Books, 1989), pp. 277-94.

26. This historical account draws on Michael H. Cardozo, "The Federal Advisory Committee Act in Operation," *Administrative Law Review*, vol. 33 (Winter 1981), pp. 1-62; Seidman and Gilmour, *Politics, Position, and Power*; The Federal Advisory Committee Act (Public Law 92-463); and *Source Book: Legislative History, Texts, and Other Documents*, Congressional Research Service Report for the Subcommittee on Energy, Nuclear Proliferation, and Federal Services of the Senate Committee on Governmental Affairs, 95 Cong. 2 sess. (GPO, July 1978) (hereafter cited as *Source Book*).

27. *Schechter Poultry Corp.* v. *United States*, 295 U.S. 495 (1935).

28. Cardozo, "FACA in Operation," p. 6.

29. *Source Book*, pp. 40-74. The bill was introduced May 9, 1957.

30. Ibid., p. 48.

31. Cardozo, "FACA in Operation," p. 8.

32. *Prescribing Regulations for the Formation and Use of Advisory Committees—Executive Order 11007*, reprinted in *Source Book*, pp. 108-10.

33. Cardozo, "FACA in Operation," p. 9. The full text is in *Source Book*, pp. 111-13.

34. See chapter 4 for a fuller discussion of this episode in the context of the early history of the EPA's Science Advisory Board.

35. Cardozo, "FACA in Operation," p. 9.

36. See *Source Book*, pp. 132-38, 141-44, 146-50, 259-90.

37. Cardozo, "FACA in Operation," p. 10.

38. Ibid.

39. Sec. 5(b) of the FACA.

40. *Lombardo* v. *Handler*, 397 F. Supp. 792 (D.D.C. 1975), aff'd 546 F.2d 1043 (D.C. Cir. 1976), cert. denied 431 U.S. 932 (1977). See also *Source Book*, p. 346.

41. Cardozo makes such an argument in "FACA in Operation," pp. 15–16. The courts have generally held against broad remedies but have left open the possibility of a court-directed restoration of balance in committee membership if objection is raised *at the time a committee is formed*, not after it has been functioning. For a summary of cases, see *Source Book*, pp. 343–48. For more recent cases, see Department of Justice, *Freedom of Information Case List* (September 1989), pp. 339–47.

42. *Nader* v. *Dunlop*, 370 F Supp. 177 (D.D.C. 1973). See also *Source Book*, p. 347.

43. *Source Book*, pp. 108–10.

44. An illustration of the problems posed by the balance requirement is seen in a case brought by Public Citizen. In *Public Citizen* v. *National Advisory Commission on Microbiological Criteria for Foods*, decided in the D.C. Circuit Court on September 26, 1989, six consumer groups challenged the composition of an advisory committee established by the U.S. Department of Agriculture to advise on food contaminants. Public Citizen, representing the consumer groups, contended that the committee's composition, mostly technical and with a majority of members from various parts of the food industry, violated the FACA because it contained no consumer representatives or public health advocates. Public Citizen lost the suit, but the splits in the three-member court illustrate the uncertainty in legal interpretation. One judge agreed with the Justice Department that plaintiffs had no standing to bring the legal action and that the FACA provided no legal remedies to enforce "fair balance" in committee membership. The other two judges rejected the Justice Department's sweeping claim of nonjudiciability but split on whether the committee was in fact balanced. One held that balance could mean the representation of different scientific fields and points of view and that under the law the agency must be given deference in interpreting its advisory needs. The other judge held that the Microbiological Committee's composition was in fact one-sided and in violation of the law. Courts generally have been unwilling to grant extensive legal remedies of the kind sought in this case, but the legal issues are far from settled.

45. See chapter 4 for a discussion of the Alar case and its conflict-of-interest aftermath.

46. Public Law 87-849 (codified as 18 U.S.C., secs. 201-18). See Bayliss Manning, *Federal Conflict of Interest Law* (Harvard University Press, 1964), for a background discussion.

47. See Roswell B. Perkins, "The New Federal Conflict of Interest Law," *Harvard Law Review*, vol. 76 (April 1963), pp. 1113–69, and, generally, Manning, *Federal Conflict of Interest Law*.

48. *Prescribing Standards of Ethical Conduct for Governmental Officers and Employees*, May 10, 1965.

49. Robert N. Roberts, *White House Ethics: The History of the Politics of Conflict of Interest Regulation* (Greenwood Press, 1988), p. 131.

50. General Accounting Office, *Action Needed to Make the Financial Disclosure System Effective*, Report to the Congress B-103987 (1977).

51. The act was amended in 1985 (Public Law 99-190, 99 Stat. 1).

52. The text of the Office of Legal Counsel, Department of Justice opinion, and an illuminating discussion of its meaning, is contained in Richard K. Berg, *Conflict-of-Interest Requirements for Federal Advisory Committees*, Report to the Administrative Conference of the United States (Washington, 1989).

53. *To Serve with Honor*, Report of the President's Commission on Federal Ethics Law Reform (Washington, 1989).

54. This account is based on interviews with officials closely involved with the FACA system during this early period and on contemporary records made available to me by officials. In particular I acknowledge the help of Robert Tarr, committee management officer of the Department of Health, Education, and Welfare, and William Bonesteel, director of the OMB Committee Management Office from 1975 to 1977. A contemporary record drawn up by Chet Warner, second director of the OMB Committee Management Secretariat, consisting of two mimeographed reports, was particularly helpful. See Chet Warner, "The Committee Management Secretariat," Report to the Senate Committee on Government Affairs (1975) (hereafter the Warner Report), and "The National Petroleum Council" (1975). I obtained these documents through the courtesy of William Bonesteel.

55. Warner Report, p. 2.

56. Ibid., pp. 3-4.

57. Ibid., pp. 5-6.

58. Ibid., pp. 6-7.

59. See Warner, "National Petroleum Council," pp. 2-5.

60. Ibid., p. 4. See also George C. Nielson, "Fuel Supplies for Winter Are Seen Adequate," *Washington Post*, September 11, 1974, p. A6; and Bob Kuttner, "New Members of Oil Council Tied to Industry," *Washington Post*, September 10, 1974, p. A4.

61. *Washington Post*, September 11, 1974, p. A6.

62. *National Anti-Hunger Coalition* v. *Executive Committee of President's Private Sector Survey on Cost Control*, 557 F. Supp. 524 (D.D.C. 1983), *aff'd* F.2d (D. Circ. 1983), *on remand* 566 F. Supp. 1515 (D.D.C. 1988).

63. Reorganization Plan No. 1, 42 Fed. Reg. 56101 (October 21, 1977). Executive Order 12024 (December 1, 1977), reprinted in *Eighteenth Annual Report of the President on Federal Advisory Committees, Fiscal Year 1989*, p. A15. Senator Lee Metcalf, chairman of the Senate Committee on Government Operations, challenged the OMB on the reasons for the transfer. Several exchanges of letters took place, but Metcalf eventually dropped the matter without a substantive response given by the OMB for the transfer.

64. For fiscal 1990 the number of advisory committees increased to 1,071

after hovering just below 1,000 for several years. *Nineteenth Annual Report of the President on Federal Advisory Committees, Fiscal Year 1990*, p. 2.

65. On the EPA crisis, see Philip B. Heymann, *The Politics of Public Management* (Yale University Press, 1987), pp. 42–49; and chapter 4 in this volume.

66. The FACA contains a provision for congressional oversight of advisory committees. The charter of each new committee according to law must be filed with the relevant oversight committees in the two houses, and thereafter, as section 5(a) stipulates, the committees are to make "a continuing review" of the activities of committees in their jurisdictions. The review should determine if the committees are to continue in existence, if their functions should be revised, and if the committees serve "a necessary function." But aside from Senator Metcalf's Subcommittee on Budgeting, Management, and Expenditure of the Senate Committee on Government Operations in the period before his death in 1978, no committee has conducted such a review. Cardozo's conclusion that "committees seem to have slight concern with their responsibilities under the Act" continues to be true today. Cardozo, "FACA in Operation," p. 19.

67. *Department of Defense/Strategic Defense Initiative Organization Compliance with Federal Advisory Committee Act*, Hearing before the Senate Committee on Governmental Affairs, 100 Cong. 2 sess. (GPO, 1988).

68. See Eric R. Glitzenstein and Patti A. Goldman, *Federal Advisory Act at the Crossroads: Needed Improvements in the Regulation of Federal Advisory Committees* (Washington: Public Citizen Litigation Group, 1989).

69. *Federal Advisory Committee Act Amendments of 1989*, Hearings before the Senate Committee on Government Operations, 101 Cong. 1 sess. (GPO, 1989).

70. See Yaron Ezrahi, "Utopian and Pragmatic Rationality: The Political Context of Scientific Advice," *Minerva*, vol. 18 (Spring 1990), esp. pp. 118–24.

Chapter Three

1. Bruce L. R. Smith, *The Rand Corporation: Case Study of a Non-Profit Advisory Corporation* (Harvard University Press, 1966).

2. The story is told in Don K. Price, *Government and Science: Their Dynamic Relation in American Democracy* (New York University Press, 1954), chap. 5; and Daniel J. Kevles, "Scientists, the Military, and the Control of Postwar Defense Research: The Case of the Research Board for National Security, 1944–46," *Technology and Culture*, vol. 16 (January 1975), pp. 20–47. A more theoretical formulation of the relation of technology to military strategy is found in Morris Janowitz, *The Professional Soldier: A Social and Political Portrait* (Glencoe, Ill.: Free Press, 1960).

3. See Price, *Government and Science*, pp. 151–52.

4. The Defense Department has a large number of other science advisory boards of potential interest that are not considered here. The DSB is my focus, but to be complete a whole series of offshoots and peripheral bodies would have to be included. Some of the other bodies omitted for present purposes are the science advisory boards of the individual services and the Strategic Defense Initiative Office (represented on the DSB by their respective chairmen), Defense Intelligence Advisory Board, DOD-University Forum, President's National Security Telecommunications Advisory Committee, Scientific Advisory Group on Effects, Public Cryptography Advisory Committee, and Graduate Medical Education Advisory Committee. At the time of the Defense Science Board's creation, by one estimate there were some 3,000 advisory committees left over from World War II that had served the Munitions Board and the predecessor to the Research and Development Board. Most of these were industry committees made up of experts on weapons and engineering applications; many probably were inactive and existed only on paper. See the testimony of Robert S. McNamara, in *Organization for National Security*, Hearings before the National Policy Machinery Subcommittee of the Senate Committee on Government Operations (the Jackson subcommittee), 86 Cong. 1 sess. (Government Printing Office, 1961), vol. 1, p. 1190. McNamara abolished some 500 committees early in his tenure and reorganized and consolidated others, but he left the DSB intact.

5. See *Hoover Commission R&D Recommendations, Department of Defense Action Program*, Progress Report, October 1956, for the steps taken by the DOD to implement the recommendations called for the previous year. I am indebted to Major Kevin R. Cunningham, assistant professor, Department of Political Science, U.S. Military Academy, for much useful information and for sharing his deep understanding of the history of the DSB and of government-science relations generally. Major Cunningham completed his doctoral dissertation on the DSB in the Department of Political Science, Stanford University, and will publish a full history of the DSB. His preliminary findings are presented in "Scientific Advisors and American Defense Policy: The Case of the Defense Science Board," paper presented at the Annual Meeting of the American Political Science Association, August 30–September 2, 1990.

6. Cunningham, "Scientific Advisors and American Defense Policy," pp. 9–14.

7. The establishment of the OSD's Advanced Research Projects Agency in early 1958 and the Defense Research and Engineering Office in late 1958 were additional steps in this complex process of reorganization. The legal framework for DOD research and engineering under the 1958 Department of Defense Reorganization Act and related legislation is summarized in "Digest of Acts and Resolutions of the 85th Congress, 2nd Session of Interest of Defense Research and Sciences," internal memorandum, Office of the Assistant Secretary of Defense/Research and Engineering, October 1958.

8. Draft charter, obtained from Major Cunningham's files.

9. Department of Defense Instruction 5128.31, December 21, 1956.

10. See Claude Witze, "Scientists Clash with Newbury on Policy," *Aviation Week*, April 15, 1957, pp. 26–27; Claude Witze, "Newbury Resigns after Six Weeks under Fire in Defense R&E Post," *Aviation Week*, April 29, 1957, p. 27; and "Power Struggles Affecting Research," *Aviation Week*, March 3, 1958, pp. 143–45. See also the discussion in Cunningham, "Science Advisors and American Defense Policy," pp. 10–12.

11. In early 1958 the two positions of assistant secretary for R&D and assistant secretary for applications engineering were combined into one position, assistant secretary for research and engineering.

12. *Minutes, Special Meeting of the Defense Science Board, 4 April 1957*, dated 10 May 1957, obtained courtesy of Major Cunningham.

13. McNamara testimony in *Organizing for National Security*, Hearings, vol. 1, p. 1190.

14. Cunningham, "Scientific Advisors and American Defense Policy," p. 16. Foster evidently was persuaded to "use them or lose them" by the DSB chairman Frederick Seitz. Foster, reportedly, was surprised to learn that he had a science advisory board.

15. See Cunningham, "Scientific Advisors and American Defense Policy," pp. 20–25.

16. Brooke Nihart, "Science Advisory Boards: Bargain or Boondoggle," *Armed Forces Journal*, March 7, 1970, p. 21.

17. I am indebted to Major Cunningham for this point.

18. See the discussion in David Flitner, Jr., *The Politics of Presidential Commissions: A Public Policy Perspective* (Transnational Publishers, 1986), pp. 187–90.

19. See "Pentagon Science Advisors Attacked," *Baltimore Sun*, July 21, 1983, p. 1.

20. *Report of the President's Blue Ribbon Commission on Defense Management*, June 1986 (known popularly as the Packard Commission).

21. The DSB completed some 260 task force reports from 1958 to 1988, with peak outputs coming under Foster in the Lyndon Johnson period, Malcolm Currie in the Ford administration, during the first Reagan term (but this was principally the culmination of work initiated under William Perry in the Carter administration), and in 1987 and 1988. See Cunningham, "Scientific Advisors and American Defense Policy," pp. 37–38.

22. Such an instance was the use of the DSB to kill the AUTODIN II data network program of the military services. In this case there was widespread dissatisfaction with the contractor's performance, but the bureaucracy was deadlocked over what to do. The DSB in effect acted as arbiter and provided the rationale for killing the program. See Cunningham, "Scientific Advisors and American Defense Policy," pp. 39–42, for a discussion of the different functions of DSB studies.

23. I am indebted to Major Cunningham for sharing his data on this point. Cunningham's unpublished data show that over the history of the DSB, the studies attracting the largest number of flag-rank participants were the 1978

study "Strategic Nuclear Balance" (the largest summer study in the DSB's history, involving sixteen flag-rank officers as advisers); the 1973 study "Electronics Management" (nine flag-rank officers involved); the 1978 study "Achieving Improved NATO Effectiveness through Armament Co-operation" (nine flag-rank officers); the 1985 study "Armor/Anti-Armor Competition" (eight flag-rank officers); and the 1986 study "Use of Commercial Components in Military Equipment" (six flag-rank or general officers).

24. Typical reports include "High Energy Lasers" (1979), "Particle Beam Weapons" (1980), and "Military Applications of New Generation Computing Technologies" (1984). Important recent studies include a review of the defense technology base (and a critique of government procurement policies), a study of the types of launch vehicles needed to sustain the military space effort (done in close collaboration with a NASA Advisory Council task force reviewing similar issues for NASA), an analysis of the close-in naval bombardment and air cover needed to support amphibious landings by the Marines, a study of the adequacy of terminal means for army control verification needs, and a critique of the technical assumptions and mission profile of the national aerospace plane program.

25. Lucy Reilly, "Cheney Aide's Proposal Seen as Slap at Betti," *Washington Technology*, December 6, 1990, p. 6.

26. Lucy Reilly, "Pentagon Elevates Tech Czar Position," *Washington Technology*, December 20, 1990, pp. 1, 6.

27. Carnegie Commission on Science, Technology, and Government, *New Thinking and American Defense Technology* (New York, August 1990), p. 20.

28. Ibid., p. 21. The task force report also contained a number of organizational recommendations to strengthen the DSB and the role of science in national security decisionmaking.

29. Elliot Marshall, "Military Labs Hit By Funding Retreat," *Science*, vol. 253 (July 12, 1991), pp. 131–32.

30. See *Authorizing Appropriations for Fiscal Years 1990. . . and for Defense Activities of the Department of Energy . . .*, House Conference Report to accompany H.R. 2461, 101 Cong. 1 sess (GPO, 1989). See also the discussion in Kevin B. Cunningham, "Scientific Advisors and American Defense Policy: The Case of the Defense Science Board," Ph.D. dissertation, Stanford University, August 1991, pp. 347–49, 414–18, 472–99; and, more generally, Arnold J. Meltsner, *Rules for Rulers: The Politics of Advice* (Temple University Press, 1990).

31. See Cunningham, "Scientific Advisors and American Defense Policy," chaps. 6, 7.

32. Quoted in ibid., pp. 478–79.

33. James Allen Smith, *The Idea Brokers: Think Tanks and the Rise of the New Policy Elite* (Free Press, 1991), pp. 231–32.

Chapter Four

1. Sheila Jasanoff, *The Fifth Branch: Science Advisors as Policymakers* (Harvard University Press, 1990). See also Allan Mazur, *The Dynamics of Technical Controversy* (Washington: Communications Press, 1981).

2. See U.S. Environmental Protection Agency, *Advisory Committees: Charters, Rosters and Accomplishments* (March 1990), for a list of the committees and their charter.

3. EPA Science Advisory Board Final Report, *The Mission and Functioning of the EPA Science Advisory Board* (October 1989), pp. 16–17. Hereafter Self-Study.

4. Self-Study, p. 4.

5. *Annual Report of the EPA Science Advisory Board*, 1989, p. A-4.

6. Ibid., p. A-1.

7. Environmental Protection Agency, *Advisory Committees*, p. 20.

8. Ibid.

9. Environmental Research, Development, and Demonstration Authorization Act of 1988, 42 U.S.C. 43G5.

10. See Joseph D. Rosen, "Much Ado about Alar," *Issues in Science and Technology*, vol. 7 (Fall 1990), pp. 85–90; R. J. Bidinotto, "The Great Apple Scare," *Reader's Digest*, vol. 137 (October 1990), pp. 53–58; and Jasanoff, *Fifth Branch*.

11. On the role of courts in environmental disputes, see H. Shep Melnick, *Regulation and the Courts: The Case of the Clean Air Act* (Brookings, 1983); David M. O'Brien, *What Process Is Due? Courts and Science-Policy Disputes* (Russell Sage Foundation, 1987); and Jasanoff, *Fifth Branch*. See also Dorothy Nelkin, *Controversy: Politics of Technical Decisions* (Beverly Hills, Calif.: Sage Publications, 1979).

12. Robert Wood, "Scientists and Politics: The Rise of an Apolitical Elite," in Robert Gilpin and Christopher Wright, eds., *Scientists and National Policy-making* (Columbia University Press, 1964), pp. 41–72.

13. These phrases are from the Clean Air Act Amendments of 1977. In weighing such requirements, the Harvard physiology professor James Whittenberger, chairman of a special EPA subcommittee on ozone, observed in 1978 that "in spite of what the legal restraints are, it seems to me that eventually we are going to have to take economics much more into consideration." See Marc K. Landy, Marc J. Roberts, and Stephen R. Thomas, *The Environmental Protection Agency: Asking the Wrong Questions* (Oxford University Press, 1990), p. 53.

14. See the analysis of Wallace S. Sayre, "Scientists and American Science Policy," in Gilpin and Wright, eds., *Scientists and National Policy-making*, pp. 97–112.

15. Self-Study, p. 6.

16. Ibid., p. 24.

17. Ibid., p. 45.

18. Ibid.

19. Landy and others, *Environmental Protection Agency*, chap. 2; Samuel P. Hays, *Beauty, Health, and Permanence: Environmental Politics in the United States, 1955–1985* (Cambridge University Press, 1987); and J. Clarence Davies III and Barbara S. Davies, *The Politics of Pollution*, 2d ed. (Indianapolis: Bobbs-Merrill, 1975).

20. See Bruce L. R. Smith, *American Science Policy since World War II* (Brookings, 1990), chap. 3.

21. Landy and others, *Environmental Protection Agency*, p. 35.

22. Ibid.

23. Ibid.

24. Ibid., p. 36.

25. Marshall R. Goodman and Margaret T. Wrightson, *Managing Regulatory Reform: The Reagan Strategy and Its Impact* (Praeger, 1987).

26. See Morton Rosenberg, *Regulating Management at OMB* (Library of Congress, American Law Division, February 1986, and postscript October 1991).

27. See Hays, *Beauty, Health, and Permanence*, pp. 337–43.

28. The 2,4,5-T episode is discussed in Perry Bush, *Uneasy Partners: A History and Analysis of the EPA Science Advisory Board* (Carnegie-Mellon University, December 1990), an administrative history of the SAB. On the earlier scientific disputes, see Robert Gilpin, *American Scientists and Nuclear Weapons Policy* (Princeton University Press, 1962); Joel Primack and Frank von Hippel, *Advice and Dissent: Scientists in the Political Arena* (Basic Books, 1974); and Mazur, *Dynamics of Technical Controversy*.

29. Constance Holden, "Critics Weigh EPA Herbicide Report, Find It Wanting," *Science*, vol. 173 (July 23, 1971), p. 312; and Nicholas Wade, "Decision on 2,4,5,-T: Leaked Reports Compel Regulatory Responsibility," *Science*, vol. 173 (August 13, 1971), pp. 610–15.

30. See *Advisory Committees*, Hearings before the Subcommittee on Intergovernmental Relations of the Senate Committee on Government Operations, 92 Cong. 1 sess. (Government Printing Office, 1971), pt. 3, pp. 755–806, especially the statement of Harrison Wellford. The words "advocate their cause" are Wellford's, not Mrak's. A reading of Wellford's testimony supports Mrak's view that the environmentalists were interested in advocacy, not analysis.

31. Wade, "Decision on 2,4,5,-T."

32. William Ruckelshaus, speech to American Chemical Society, September 13, 1971, quoted in *Advisory Committees*, Hearings, pt. 3, p. 833.

33. EPA, "Rosters and Functions for EPA Public Advisory Committees" (October 1971), reproduced in *Advisory Committees*, Hearings, pt. 3, pp. 848–76.

34. *Advisory Committees*, Hearings, pt. 3, p. 832.

35. I had a voice mail exchange with Victor Reinemer, a close Metcalf

aide, during this period that supported this version of Metcalf's thinking. Unfortunately, Reinemer died suddenly before I had a chance for further discussion and before he had a chance to comment on a draft of the manuscript.

36. Hays, *Beauty, Health, and Permanence*, pp. 348–49.

37. Stanley Greenfield, "Information Memorandum," July 19, 1972, quoted in Bush, *Uneasy Partners*, p. 12.

38. Ibid.

39. Stanley M. Greenfield, "Action Memorandum," August 15, 1972, quoted in Bush, *Uneasy Partners*, p. 13.

40. Bush, *Uneasy Partners*, p. 13.

41. Crandall and Lave, perhaps overstating their case slightly, argue that political factors, not scientific or legal factors, predominate in decisionmaking on health, safety, and environmental issues. See Robert W. Crandall and Lester B. Lave, eds., *The Scientific Basis of Health and Safety Regulation* (Brookings, 1981), pp. 1–17.

42. Interview with Stanley M. Greenfield, in SAB files, cited in Bush, *Uneasy Partners*, pp. 14–15.

43. Greenfield, "Action Memorandum," cited in Bush, *Uneasy Partners*, pp. 15–16. Ruckelshaus meanwhile had moved to the Justice Department to serve as deputy attorney general under Elliot Richardson.

44. "Train Urges Closer Cooperation with Scientific Community," *Environmental News*, January 27, 1974, cited in Bush, *Uneasy Partners*, p. 17.

45. Alvin Alm, "Action Memorandum," September 10, 1974, cited in Bush, *Uneasy Partners*, pp. 17–18.

46. Interview with Roger McClellan, cited in Bush, *Uneasy Partners*, p. 20.

47. See Jasanoff, *Fifth Branch*, chap. 4, for an illuminating analysis of this problem.

48. Ibid., pp. 86–87.

49. Ibid., p. 86.

50. Quoted in ibid., p. 87.

51. Environmental Research, Development, and Demonstration Authorization Act (ERDDAA) of 1978, 42 U.S.C. 4365.

52. Jasanoff, *Fifth Branch*, p. 87.

53. See Goodman and Wrightson, *Managing Regulatory Reform*.

54. General Accounting Office, *Attrition of Scientists and Engineers at Seven Agencies*, Report to the Congress B-209389 (May 29, 1984), p. 5.

55. Confidential interview with the author.

56. Jasanoff, *Fifth Branch*, p. 88.

57. Ibid., p. 89.

58. National Academy of Sciences–National Research Council, *Risk Assessment in the Federal Government: Managing the Process* (Washington: National Academy Press, 1983).

59. See Jasanoff, *Fifth Branch*, chap. 9.

. 60. Sheila Jasanoff, *Risk Management and Political Culture* (New York: Russell Sage Foundation, 1986), pp. 41–53.

61. Landy and others, *Environmental Protection Agency.*

62. See Leslie Roberts, "Counting on Science at EPA," *Science*, vol. 249 (August 10, 1990), pp. 616–18, for a description of the effort.

63. Ibid., p. 616. See also Richard Morgenstern and Stuart Sessions, "Weighing Environmental Risks: EPA's Unfinished Business," *Environment*, vol. 30 (July–August 1988), pp. 14–17.

64. Jasanoff, *Fifth Branch*, p. 96.

65. U.S. Environmental Protection Agency, Science Advisory Board, *Future Risk: Research Strategies for the 1990's* (September 1988). Information on distribution acquired from the SAB.

66. Jasanoff, *Fifth Branch*, p. 96.

67. William K. Reilly, "Aiming before We Shoot: The Quiet Revolution in Environmental Policy," speech to the National Press Club, September 26, 1990.

68. Bruce N. Ames and Lois Swirsky Gold, "Too Many Rodent Carcinogens: Mitogenesis Increases Mutagenesis," *Science*, August 31, 1990, pp. 970–71.

69. 104 Stat. 2399, passed on November 15, 1990.

70. Jasanoff, *Fifth Branch*, p. 84.

71. On the role of the chairman generally in successful advisory committee operations, see Richard A. Wegman, *The Utilization and Management of Federal Advisory Committees*, Report of the Charles F. Kettering Foundation (New York, 1982), pp. 92–102.

Chapter Five

1. For the history, see Harold P. Green and Alan Rosenthal, *Government of the Atom: The Integration of Powers* (Atherton Press, 1963).

2. *Replies from Executive Departments and Federal Agencies (January 1, 1953–January 1, 1956)*, Committee Print, House Committee on Government Operations, 84 Cong. 2 sess. (Government Printing Office, 1956), pt. 7, pp. 2984–3025.

3. Letter now in the files of Joel Snow, former senior Energy official and now associate director, Argonne National Laboratory. I am indebted to Dr. Snow for sharing his recollections of these events.

4. U.S. Department of Energy, Energy Research Advisory Board, *Report of the Energy Research Advisory Board on the Relationship between the University of California and the Los Alamos Scientific and the Lawrence Livermore Laboratories* (Washington, 1979).

5. The regents did decide in a compromise gesture to create an Institute for the Study of Global Conflict. The institute would study arms control and seek to allay the fears of those who felt that affiliation with weapons labora-

tories placed the university in a position of supporting the arms race. Herbert York became the first director of the institute, which was located at the University of California at San Diego.

6. See Bruce L. R. Smith, *American Science Policy since World War II* (Brookings, 1990), chap. 4.

7. ERAB reports completed from November 1979 to November 1980 include those on construction-development project management, evaluation of the proposed coal gasification multitest facility, gasohol review, geothermal panel on high temperature resources development, study group on hot dry rock, fusion review panel on DOE's magnetic fusion program, advanced isotope separation, and the panel on R&D needs in the department.

8. Data obtained from internal ERAB materials. For an assessment of the impact of these studies on departmental policy, see Argonne National Laboratory, *Case Studies of DOE Response to ERAB Recommendations: A Retrospective Analysis* (Chicago, 1986).

9. The situation was mitigated, and the transition smoothed, by the close friendships of Secretary of Energy Donald P. Hodell (1982–85) with ERAB chairman Ralph Gens, and Secretary of Energy John S. Herrington (1985–89) with ERAB chairman John Schoettler, who had been his college roommate.

10. NAS Committee to Assess Safety and Technology Issues at DOE Reactors, "Safety Issues at the Defense Production Reactors: A Report to the Dept. of Energy" (National Academy Press, 1987).

11. As of 1990 there were fourteen official FACA-chartered advisory committees in the Department of Energy.

14. Argonne National Laboratory, *Case Studies of DOE Response to ERAB Recommendations*. Meeting of the Energy Research Advisory Board, August 1986.

15. William T. Gormley, *Taming the Bureaucracy: Muscles, Prayers, and Other Strategies* (Princeton University Press, 1989), pp. 74–77.

16. Department of Energy, Fusion Policy Advisory Committee, *Final Report* (September 1990).

Chapter Six

1. For another example of this support function, the experience of the Forest Service is instructive. See R. D. Forbes, memorandum on advisory council, October 9, 1954, quoted in Ashley L. Schiff, *Fire and Water: Scientific Heresy in the Forest Service* (Harvard University Press, 1962), p. 166. Also, the early history of the Naval Research Advisory Committee (NRAC), illustrates a related purpose: it is an example of a committee whose only purpose was to lobby for the research budget of its parent agency, the Office of Naval Research. See Harvey M. Sapolsky, *Science and the Navy: The History of the Office of Naval Research* (Princeton University Press, 1990), chap. 6, esp. pp. 102–07.

2. The mixed fleet study was done under a task force chaired by General Jasper Welch, USAF, retired. See L. J. DeRyder, ed., *Assessment of Mixed Fleet Potential for Space Station Launch and Assembly* (National Aeronautics and Space Administration, 1987).

3. Observing a recent NASA Advisory Council meeting, I had the impression that a presentation by General Jasper Welch engaged the active interest of NASA officials and of the council members. He made proposals for example to "spend money at the front end [that is, on development and on the upgrading of vehicle design] to save money on operational costs later."

The presentation sparked a lively and informal discussion. Agency officials responded and defended their approach; the discussion was substantive. Other presentations at the meeting wandered off into meandering, descriptive briefings by NASA officials of their programs. The contrast between the parts of the meeting was striking. Press and media attendance thinned down as the more routine briefings droned on; few questions were asked. The difference between an advisory committee that "matters" and one that does not is illustrated in microcosm in this episode. Sometimes a committee can engage the attention of officials; at other times it is little more than a ceremonial interruption of the day's activity.

General Welch's aplomb in presenting his committee's findings was further shown when, as he rose to speak, his trousers split at the seam. I was sitting in the row directly behind him and was perhaps one of the few to notice. With a graceful side movement he took his place behind the podium and launched into his talk.

4. *America at the Threshold*, Report of the Synthesis Group on America's Space Exploration Initiative (Arlington, Va., 1991). For a commentary, see Gene Koprowski, "Synthesis Panel Puts Man on Mars in 2014," *Washington Technology*, June 13, 1991, p. 1.

5. Committee on Space Science and Applications, *The Crisis in Space and Earth Science* (Washington, 1986).

6. National Research Council Aeronautics and Space Engineering Board, *Report of the Committee on a Commercially-Developed Space Facility* (Washington, 1989). Strong opposition from the aerospace industry, and from congressional supporters of the space station, helped to sink the CDSF proposal, which was perceived to be a threat to funding for the space station.

7. *Leadership and America's Future in Space: A Report to the Administrator* (NASA, 1987).

8. On Reagan's science policies, see Bruce L. G. Smith, *American Science Policy since World War II* (Brookings, 1990), chap. 6.

9. National Academy of Science Committee on Human Exploration of Space, "Human Exploration of Space: A Review of NASA's 90-Day Study and Alternatives" (National Academy Press, 1990).

10. Daniel Fink and Louis Lanzerotti became members of the Advisory Committee on the Future of the U.S. Space Program. Later, after it had concluded its work, committee member and former Xerox chief executive

David Kearns was appointed to the NASA Advisory Council. (He resigned when later nominated as deputy secretary of education by President Bush.)

11. Robert L. Park, "Say No to This 'Orbiting Pork Barrel,' " *Washington Post*, July 10, 1991, p. A21.

12. Ibid.

Chapter Seven

1. Quoted in Carnegie Commission on Science, Technology, and Government, *Science and Technology in U.S. International Affairs*, Report of a Steering Group, Rodney W. Nichols, chairman (New York, 1992), p. 32.

2. Ibid. For a history of science in the State Department, see ibid., chap. 3, and House Committee on International Relations, *Science, Technology and American Diplomacy*, Committee Print (Government Printing Office, 1977), vol. 2, chap. 14.

3. See Lincoln P. Bloomfield, "Planning Foreign Policy: Can It Be Done?" *Political Science Quarterly*, vol. 93 (Fall 1978), pp. 369–91.

4. The late Frank Huddle of the Congressional Research Service was an active force in identifying what he believed were weaknesses in the United States' approach to science and technology in foreign affairs and was a primary force behind enactment of title V, section 504(a) in the authorization act for the department.

5. A report on science and foreign policy operations, done by a committee under the chairmanship of T. Keith Glennan, former NASA administrator, had recommended such a step. *Technology and Foreign Affairs*, a report by Dr. T. Keith Glennan to Deputy Secretary of State Charles W. Robinson (Department of State, December 1976). The Glennan committee is an interesting example of an informal effort to provide science advice to State Department policymakers. The report was prepared not through a formal advisory committee but through a cluster of consultants acting as a committee to avoid the formalities required under the Federal Advisory Committee Act (FACA) and to avoid incurring the wrath of the OMB, which was bent on pruning the number of committees. A review of the B-1 bomber in the DOD in the early part of the Carter administration was done in a similar fashion, as described in Herbert York's engrossing memoir, *Making Weapons, Talking Peace: A Physicist's Odyssey from Hiroshima to Geneva* (Basic Books, 1987), pp. 265–67. Anyone who sought to create a committee during this period got the impression that the FACA was burdensome, mainly because of the OMB's attitude at the time (which reflected the president's wishes). Even though it later became easier to establish committees under the FACA, the impression lingered in the minds of agency officials.

6. This breakdown was obtained by studying Pickering's calendar over a six-week period, informally classifying meetings under the headings of *nuclear, fisheries*, and so on, and was done at my direction during my service as planning deputy. A different breakdown is by the areas in which science and

technology agreements have been negotiated. See Carnegie Commission, *Science and Technology in U.S. International Affairs*, p. 56.

7. The OMB posture, as described in chapters 4 and 6, prompted the consolidation of two NASA committees to form the NASA Advisory Council and the move of the EPA Science Advisory Board to a statutory base (as a way to escape constant OMB pressure to cut committees). For the general history, see chapter 2.

8. Committee members, especially those from the West Coast, found committee membership burdensome and attendance suffered as a result.

9. Special ambassadors were in charge of other policy matters within OES jurisdiction, for example, ambassador Jean Wilkonski headed the delegation for the United Nations Conference on Science and Technology for Development. She drew on the bureau for staff support. This State Department tradition of the special ambassador has many of the same advantages and disadvantages of matrix management within the business firm.

10. For a witty and satirical assault on the Law of the Sea negotiations, see Charles Horner, "Two More If by Sea," *American Spectator*, vol. 23 (August 1990), pp. 15-17. Horner served in the OES as deputy assistant secretary for science and technology from 1982 to 1985.

11. For proposals on how to improve the recruitment process for scientists, the technical training of foreign service officers, and the use of scientists in State Department operations generally, see Carnegie Commission, *Science and Technology in U.S. International Affairs*, pp. 9-16, 83-100.

Chapter Eight

1. See William T. Golden, ed., *Science Advice to the President* (Pergamon, 1980), *Science and Technology Advice to the President, Congress, and Judiciary* (Pergamon, 1988), and *Worldwide Science and Technology Advice to the Highest Levels of Government* (Pergamon, 1991); Bruce L. R. Smith, *American Science Policy since World War II* (Brookings, 1990), pp. 117-19; and Carnegie Commission on Science, Technology, and Government, *Science & Technology and the President* (New York, 1988).

2. A minor indicator is that neither President Jimmy Carter nor President Ronald Reagan made any reference to their science advisers in their published memoirs. See Jimmy Carter, *Keeping Faith: Memoirs of a President* (Bantam Books, 1982); and Ronald Reagan, *An American Life* (Simon and Schuster, 1990).

3. Edward Burger, *Science at the White House: A Political Liability* (Johns Hopkins University Press, 1980), pp. 9, 115. See also David Z. Beckler, "The Precarious Life of Science in the White House," *Daedalus*, vol. 103 (Summer 1974), pp. 115-34.

4. Lewis E. Auerbach, "Scientists in the New Deal: A Pre-War Episode in the Relations between Science and Government in the United States," *Minerva*, vol. 3 (Summer 1965), p. 467.

5. See Smith, *American Science Policy since World War II*, chap. 2.

6. The story is well told in a number of important works. See, among others, Richard G. Hewlett and Oscar E. Anderson, *A History of the United States Atomic Energy Commission* (Pennsylvania State University Press, 1962); Daniel J. Kevles, *The Physicists* (Knopf, 1978); Richard Rhodes, *The Making of the Atomic Bomb* (Simon and Schuster, 1989); and Robert Jungk, *Brighter Than a Thousand Suns* (Harcourt Brace, 1958).

7. *Report of the President's Committee on Administrative Management* (the Brownlow committee) (Government Printing Office, 1937).

8. See the new release of the Bush Report with an introduction by Daniel J. Kevles, *Science: The Endless Frontier* (Washington, 1990); and Smith, *American Science Policy since World War II*, chap. 3. I have also benefited from a discussion with Oscar Ruebhausen, who acted as liaison between Cox and Bush in the negotiations leading to the Bush Report.

9. Golden, ed., *Science Advice to the President*, pp. 245–56. See also Smith, *American Science Policy since World War II*, pp. 111–19. A detailed and carefully documented account of scientists serving as advisers at the White House level is Gregg Herken, *Cardinal Choices: The Presidents' Science Advisers from Roosevelt to Reagan* (Oxford University Press, 1992).

10. Observation made by David Z. Beckler at a conference on presidential science advising held by the University of California, San Diego, in La Jolla, California, March 1988. For the conference proceedings, see *Science, Technology, and Government: A Crisis of Purpose?* Richard C. Atkinson, chairman (University of California at San Diego, 1989).

11. George B. Kistiakowsky, who succeeded Killian as special assistant in 1959, was somewhat more focused on policy *for* science than Killian, although only one formal PSAC report focused on academic science in his term of service. David Z. Robinson, who served on the Office of Science and Technology staff from 1961 to 1967, estimated in private correspondence that policy *for* science constituted only about 5 percent of the time of the OST and the presidential science adviser during his period of service. The shift described in the text occurred somewhat later, but probably never exceeded 10 percent of the adviser's effort.

12. I served on the staff of the OST and the Budget Bureau in 1968, and in the course of my work I interviewed a number of President Johnson's close advisers and attended numerous staff meetings. This account draws on my firsthand experience.

13. Thaddeus Trenn considers 1965 the turning point, the date marking the loss of influence in national security policy. See Thaddeus J. Trenn, *America's Golden Bough: The Science Advisory Intertwist* (Oelgeschlager, Dunn, and Hain, 1983), chap. 4. That the OST had little significant involvement in national security policy was apparent to me in 1968.

14. See Smith, *American Science Policy since World War II*, chap. 4.

15. Frederick Seitz, "Matching Reality to Need," in Golden, ed., *Scientific Advice to the President, Congress, and Judiciary*, p. 322.

16. For a full list of OST's achievements and of PSAC reports, see

Beckler, "The Precarious Life of Science in the White House," pp. 125–27, 132–34.

17. See Burger, *Science at the White House*, pp. 77–83.

18. Nixon later backed off from his opposition to Long, but the latter declined the post. The "Long incident" was something of a *cause célèbre* in the scientific community. See Philip M. Buffey, "Nixon and NSF: Politics Block Appointment of Long as Director," *Science*, vol. 16 (April 18, 1969), p. 283; and the discussion in Herken, *Cardinal Choices*, pp. 166–68.

19. S. J. Buchsbaum, "On Advising the Federal Government," in Golden, ed., *Scientific Advice to the President, Congress, and Judiciary*, p. 71.

20. Ibid., pp. 71–72.

21. As Don K. Price summarized the matter then, "The potential strains of the Vietnam war period were too great, and the formal abolition of the PSAC was accepted by Congress with hardly a word of protest. Obviously, the rift between the political views of the Nixon (and for that matter of the Johnson) administration was too great to permit the continuation of so intimate and confidential a relationship. See Price, "Money and Influence: The Links of Science to Public Policy," *Daedalus*, vol. 103 (Summer 1974), p. 109.

22. See Herken, *Cardinal Choices*, pp. 168–70.

23. William Chapman, "Nixon Science Unit Cites ABM Flaws," *Washington Post*, March 14, 1969, p. A1.

24. Kissinger, *White House Years*, p. 540.

25. Golden, ed., *Science Advice to the President*, p. 123.

26. *Department of Transportation and Related Agencies' Appropriations for 1971*, Hearings before a Subcommittee of the House Committee of Appropriations, 91 Cong. 2 sess. (GPO, 1990), pt. 3, pp. 980–94; *Economic Analysis and the Efficiency of Government*, Hearings before the Subcommittee on Economy and Government of the Joint Economic Committee, 91 Cong. 2 sess. (GPO, 1970), pt. 4, pp. 904–20; and *Department of Transportation and Related Agencies' Appropriations*, Hearings before the Senate Committee on Appropriations, 91 Cong. 2 sess. (GPO, 1970), pp. 1621–81.

27. William G. Wells, Jr., "Science Advice and the Presidency: An Overview from Roosevelt to Ford," Ph.D. dissertation, University of Michigan, December 1977, pt. 2, p. 650; and Herken, *Cardinal Choices*, pp. 117–81.

28. Beckler, "Precarious Life of Science in the White House," p. 116.

29. Stever was sworn in on August 12, 1976, as the science adviser. The position had been recreated in the National Science and Technology Policy, Organization, and Priorities Act of 1976, Public Law 94-282, signed May 11, 1976, by President Ford (H.R. 10230).

30. Price, "Money and Influence," p. 106. See also Victor F. Weisskopf, "The Significance of Science," *Science*, vol. 176 (Spring 1972), pp. 138–46.

31. See Hedrick Smith, *The Power Game: How Washington Works* (Ballantine Books, 1988), pp. 601–05.

32. A report that helped shape his thinking was *Science & Technology and*

the President, A Report of the Carnegie Commission on Science, Technology, and Government (New York, 1988).

33. For a discussion, see ibid., pp. 58–71.

34. Executive Office of the President, Office of Science and Technology Policy, *U.S. Technology Policy* (Washington, 1990).

35. See Martha Derthick, *On Commissionship—Presidential Variety* (Brookings, 1972); and David Flitner, Jr., *The Politics of Presidential Commissions* (Transnational Publishing, 1986).

36. President's Commission on Industrial Competitiveness, *Global Competition: The New Reality*, vols. 1 and 2 (Washington, 1985).

37. *Report of the Presidential Commission on the Human Immunodeficiency Virus Epidemic* (Washington, 1988).

38. *Public Citizen* v. *National Economic Commission*, Civil Action 88-3484, December 1988.

39. See Paul C. Light, *Artful Work: The Politics of Social Security Reform* (Random House, 1985).

40. See Flitner, *Politics of Presidential Commissions*, pp. 179–81.

41. Elizabeth Drew could write in 1968 at a high point in the creation of presidential commissions that "the technique of appointing a special presidential commission . . . to investigate, obfuscate, resolve, defuse, defer, detail or derail a problem has become as much an instrument of the presidency as the State of the Union Message, the toss of the ball on opening day, or the review of troops in wartime." Such an assessment would not ring true today. Elizabeth Drew, "On Giving Oneself a Hotfoot: Government by Commission," *Atlantic*, vol. 221 (May 1968), p. 45.

42. GSA Committee Management Secretariat, *Eighteenth Annual Report of the President on Federal Advisory Committees, Fiscal Year 1989*, p. 33.

43. *Presidential Advisory Committees*, Hearing before a subcommittee of the House Committee on Government Operations, 91 Cong. 2 sess. (GPO, 1970), vol. 2, p. 33.

44. Amos A. Jorden, William J. Taylor, and Lawrence J. Korb, *American National Security: Policy and Process*, 3d ed. (Johns Hopkins University Press, 1989), pp. 138–41.

45. The Federal Employees Pay Comparability Act of 1990, incorporated into and officially cited as Treasury, Postal and General Appropriations Act, 1991, Public Law 101-509, passed November 5, 1990.

46. Arnold Meltsner, *Rules for Rulers* (Temple University Press, 1990), p. 128.

47. Don K. Price, *The Scientific Estate* (Belknap Press of Harvard University Press, 1965), chap. 7.

48. Yehezkel Dror, "Conclusions," in William Plowden, ed., *Advising the Rulers* (Oxford: Blackwell, 1987), pp. 185–215. See also Dror, *Public Policymaking Reexamined* (Chandler Publishing, 1968).

49. Alexander L. George, *Presidential Decision-Making in Foreign Policy:*

The Effective Use of Information and Advice (Boulder, Colo.: Westview Press, 1980).

50. Irving L. Janis, *Groupthink*, 2d ed. (Houghton Mifflin, 1982).

Chapter Nine

1. James Allen Smith, *The Idea Brokers: Think Tanks and the Rise of the New Policy Elite* (Free Press, 1991), p. 238.

2. The "pure" science agencies, perhaps because they have so little independent political power, have been most vulnerable to patronage appointments in their advisory ranks (excluding grant review committees, which have remained nonpolitical). Since the late 1970s the National Science Board has been composed of people representing constituencies rather than people chosen primarily for their scientific distinction. The advisory councils of the various national institutes of health have been a favorite source of patronage appointments.

3. Frederick Seitz, "Matching Reality to Need," in William T. Golden, ed., *Science and Technology Advice to the President, Congress, and Judiciary* (Pergamon Press, 1988), p. 324.

4. Robert Wood, "The Rise of an Apolitical Elite," in Robert Gilpin and Christopher Wright, eds., *Science and National Policy-making* (Columbia University, 1964), pp. 41–72; and Yaron Ezrahi, *The Descent of Icarus* (Harvard University Press, 1990).

5. Smith, *Idea Brokers*, chap. 10.

6. See Sheila Jasanoff, *The Fifth Branch* (Harvard University Press, 1990), chaps. 10, 11.

7. Joel Doble and Jean Johnson, *Science and the Public: A Report in Three Volumes*, Study for the Charles F. Kettering Foundation Comparing the Perspectives of Citizens and Scientists (New York: Kettering Foundation, 1990).

8. Doble and Johnson, *Science and the Public*, vol. 1: *Searching for Common Ground on Issues Related to Science and Technology*, p. ix.

9. Ibid., p. xi.

10. An example of a formal consultative procedure is the Prospective Payment Assessment Commission (ProPAC) of the Department of Health and Human Services (HHS). ProPAC, an independent advisory commission serving both HHS and Congress, recommends updates and modifications in the hospital reimbursement rates under the prospective payment reimbursement system adopted in the 1985 Social Security Act Amendments (Public Law 98-21, 97 stat. 63). The Health Care Financing Administration (HCFA) is obligated to consider the ProPAC's annual recommendations before setting final reimbursement rates. The HCFA responds according to an elaborate and formal procedure, including publication in the Federal Register, but rarely modifies its final rule on the basis of the ProPAC advice. See Prospective Payment Assessment Commission, *Report and Recommendations to the Secretary, U.S. Department of Health and Human Services* (Washington, 1983).

Index

Abbington, David, 61
Administrative Conference of the United States, 199
Advanced Research Projects Agency. *See* Defense Advanced Research Projects Agency
Advisory Committee on Nuclear Facility Safety, 110
Advisory Committee on the Future of the U.S. Space Program. *See* Augustine committee
Advisory committees, 3–4, 81; under Carter administration, 40–42; Congress and, 43–44; criticism of, 30, 31; defined by FACA, 26–27; in eighteenth and nineteenth centuries, 14–16; membership, 9–10, 26–32; number of, 40; potential conflicts of interest, 31–36; purpose and use of, 5–9, 35–36, 129–31; under Reagan administration, 42–43; regulatory framework, 21. *See also* Federal Advisory Committee Act of *1972*; Science advisory committees
Advisory Panel on Ethics and Conflicts of Interest in Government, 32
Agriculture, U.S. Department of, 24, 81
Alar, 72
American Association for the Advancement of Science, 92, 132
American Statistical Association Committee on Energy Statistics, 110
Ames, Bruce, 97
Antiballistic Missile (ABM) program, 171–73, 177
Argonne National Laboratory, 112, 113

Arms Control and Disarmament Agency (ACDA), 166, 167
Ash, Roy, 75
Ash Council 75, 76, 176
Atomic Energy Act of *1946*, 102
Atomic Energy Commission, 102, 103, 162, 163
Auerbach, Lewis E., 161
Augustine committee, 132, 133–36, 160, 183

Baker, Howard, 139
Baker, James A., *III*, 150
Basic Energy Sciences Advisory Committee, 110
Bath, Thomas, 85, 86, 87
Beckler, David, 176
Beggs, James, 174
Benedict, Richard, 150
Berkner, Lloyd V., 138
Berkner Report, 138
Bernthal, Frederick, 149
Betti, John, 61
Bohlen, Curtis, 152
Bonesteel, William, 40
Bourne, Peter, 40
Bribery, Graft, Conflicts of Interest Act of *1962*, 32
Bromley, D. Allan, 179–80, 181
Brown, Harold, 54, 55, 63
Brown, Jerry, 104
B-2 bomber, 63, 64
Buchsbaum, Solomon J., 55, 104, 105, 181
Buckley, Oliver, 163
Bundy, McGeorge, 166

Office of Science and Technology Policy
    (OSTP), 178
Office of Scientific Research and Develop-
    ment (OSRD), 49, 155
Office of Senate Legal Counsel, 33
Office of the Secretary of Defense (OSD),
    50, 51, 53, 56, 59

Packard, David, 57
Packard commission, 57–59
Partial Test Ban Treaty of *1962*, 166–67
Pell, Claiborne, 139
Perry, William J., 61
Pickering, Thomas R., 141, 143, 145, 147,
    153
Policymakers: at OMB, 77–78; process,
    46–47, 48; 2,4,5,-T herbicide con-
    troversy and, 79–82. *See also* Scien-
    tific advisers and agency
    policymakers
Pollack, Herman, 138, 139, 141
Powell, Adam Clayton, Jr., 33
Presidential science advisers and advisory
    committees, 17, 182, 184, 188; back-
    ground, 160–61, 172; under Bush
    administration, 179–82; under Carter
    administration, 178; under Eisen-
    hower administration, 163–66, 176;
    under Ford administration, 177; un-
    der Johnson administration, 160,
    167–68, 176; under Kennedy admin-
    istration, 166–67, 176; under Nixon
    administration, 160, 169–70, 171–78;
    under Reagan administration, 178–
    79; role of, 155, 156–57,
    158–59, 178; under Roosevelt ad-
    ministration, 161–62, 172; science
    advisers amd agency policymakers,
    171–72, 185–88; under Truman ad-
    ministration, 162–63; 2,4,5,-T herbi-
    cide controversy, 170. *See also*
    Office of Science and Technology;
    Office of Science and Technology
    Policy; President's Council of Ad-
    visers on Science and Technology;
    President's Science Advisory Com-
    mittee
President's Commission on Ethics Law
    Reform, 34
President's Committee on Administrative

Management (Brownlow committee),
    161–62, 163
President's Council of Advisers on Science
    and Technology (PCAST), 179,
    180–82
President's Foreign Intelligence Advisory
    Board, 184
President's Science Advisory Committee
    (PSAC), 51, 159–60; under Eisen-
    hower administration, 180; under
    Johnson administration, 168–69;
    under Nixon administration, 169–70,
    171–78; origin, 164–65
Press, Frank, 92, 105, 131, 178
Priestley, Joseph, 12
Public Citizen, 30–31, 43–44

Quarles, Donald, 53
Quarles, John, 83–84

Rabi, I. I., 163
Ramo, Simon, 139
Ray, Dixie Lee, 140, 146, 147, 154
Reagan administration, 31, 77; DSB and,
    56–57; ERAB and, 106–9; NASA
    and, 132; presidential science advis-
    ers, 178–79; SAB and, 42, 68, 74,
    89–92, 98–100; State Department
    science advisers and, 146, 147–50
Regulatory Analysis Review Group, 77
Regulatory policy, 89–92, 99, 202
Regulatory Reform Task Force, 77
Reilly, William, 70, 74, 95, 96–97, 100
Research and development, 165, 179;
    DOD, 49, 50–53, 57, 59, 60, 61–62;
    EPA, 77, 96; DOE, 110, 116, 120;
    OST, 176; post–World War II re-
    forms, 19–20
Research and Development Board (RDB),
    49, 50
Robinson, Charles, 143
Roddis, Lewis H., Jr., 107
Roosevelt administration: presidential sci-
    ence advisers, 161–62, 172
Roth, William, 44
Ruckelshaus, William D.: as EPA admin-
    istrator, 24, 76–82; 92–95; position
    on SAB, 84–85
Rusk, Dean, 138, 139